François Cardarelli

Sulfuric Acid Digestion, Sulfuric Acid Baking, and Sulfation Roasting in Mineral and Chemical Processing, and Extractive Metallurgy

Electrochem Technologies & Materials Inc.

François Cardarelli, President & Owner
Electrochem Technologies & Materials Inc.
2037 Aird Avenue, Suite 201
Montréal (QC) H1V 2V9, Canada
www.electrochem-technologies.com
Member of ACS, AIChE, CIC, ECS, MSA, OCQ, OS, and TMS

Email: contact@electrochem-technologies.com
          contact@francoiscardarelli.ca

ISBN 978-1-7775769-0-5 (Softcover)

ISBN 978-1-7775769-1-2 (eBook)

Printed on acid-free paper.

The Electrochem Technologies & Materials Inc. imprint is published by the Canadian corporation Electrochem Technologies & Materials Inc.

The registered company address is: 2037 Aird Avenue, Suite 201, Montréal (Qc) H1V 2V9, Canada

Units Policy

In this monograph all the units of measure used for describing physical quantities and properties were those recommended by the *Système International d'Unités* (SI) except in some few instances where some units from the *US Customary System* (USCS) are used in conjunction. For accurate conversion factors between these units and the other non SI units (e.g., cgs, fps, Imperial, and US customary systems) please refer to the reference book of the same author:

CARDARELLI, F. (2005) *Encyclopaedia of Scientific Units, Weight and Measures. Their SI Equivalences and Origins*. Springer, New York, London, xxiv, 848 pages; ISBN 978-1-85233-682-0.

Books by the same author:

CARDARELLI, F. (2018) *Materials Handbook. A Concise Desktop Reference, Third edition*. Springer, Cham, London, New York. cxxxii, 2254 pages, 150 black and white figures, 25 illustrations in color in two volumes. ISBN 978-3-319-38923-3.

CARDARELLI, F. (2008) *Materials Handbook. A Concise Desktop Reference, Second edition*. Springer, London, xxxviii, 1340 pages, 55 figures; ISBN 978-1-84628-689 (Out-of-Print).

CARDARELLI, F. (2005) *Encyclopaedia of Scientific Units, Weight and Measures. Their SI Equivalences and Origins*. Springer, New York, London, xxiv, 848 pages; ISBN 978-1-85233-682-0.

CARDARELLI, F. (2000) *Materials Handbook. A Concise Desktop Reference*. Springer, London, New York, ix, 595 pages, 7 figures; ISBN 978-1-85233-168-2 (Out-of-Print).

CARDARELLI, F. (1999) *Scientific Unit Conversion. A Practical Guide to Metrication, Second edition*. Springer, London, New York, xvi, 488 pages; ISBN 978-1-85233-043-0 (Out-of-Print).

CARDARELLI, F. (1997) *Scientific Unit Conversion. A Practical Guide to Metrication*. Springer, London, Heidelberg, xvi, 456 pages, ISBN 978-3-540-76022-9 (Out-of-Print).

Author Biography

François Cardarelli
Canadian & French citizen

Academic Background

Ph.D. (Doctorat), Chemical Engineering (Université Paul Sabatier, Toulouse, 1996)

DEA in Electrochemistry (Université Pierre et Marie Curie, Paris, 1992)

M.Sc. (Maîtrise), Physical Chemistry (Université Pierre et Marie Curie, Paris, 1991)

B.Sc. (Licence), Physical Chemistry (Université Pierre et Marie Curie, Paris, 1990)

DEST (Credits) in Nuclear Sciences and Technologies (CNAM, Paris, 1988)

DEUG B in Geophysics and Geology (Université Pierre et Marie Curie, Paris, 1987)

Baccalaureate C (Mathematics, Physics, and Chemistry) (CNED, Versailles, France, 1985)

Working Areas

The author has worked in the following areas since 1990 until present:

2010-Present President and owner, Electrochem Technologies & Materials Inc., Montreal (Quebec), Canada, manufacturing industrial electrodes and electrolyzers, producing vanadium, tantalum, and tungsten chemicals, inventing, patenting, and commercializing electrochemical, chemical, and metallurgical processes for recycling by-products.

2008-2010 Recycling manager, 5N Plus Inc., Ville Saint-Laurent (Quebec), Canada, in charge of the recycling of end-of-life cadmium telluride (CdTe) thin-film photovoltaic solar panels and the hydrometallurgical recovery of tellurium and cadmium.

2007-2008 Principal electrochemist, Materials and Electrochemical Research (MER) Corp., Tuscon (Arizona), USA, working on the electrowinning of titanium metal powder from composite Ti_2OC anodes in molten salts, and other materials-related projects.

2000-2007 Principal chemist (materials), technology department, Quebec Iron and Titanium (QIT) now Rio Tinto, Sorel-Tracy (Quebec), Canada, invented the electrowinning of titanium metal from molten titanium slags and on other novel electrochemical processes.

1998-200 Materials expert and industrial electrochemist, lithium department, Avestor (now Blue Solutions), involved in the metallurgy and processing of lithium metal anodes and the recycling of spent lithium metal polymer batteries.

1997-1998 Battery product leader, technology department, Argotech Productions, Inc. (Avestor), Boucherville (Québec), Canada, in charge of electric-vehicle, stationary, and down-hole oil-drilling applications of lithium metal polymer batteries.

1996-1997 Registered consultant in chemical and electrochemical engineering (Toulouse, France) providing scientific advices on electrochemical processes and electrode materials.

1993-1996 Research scientist, Laboratory of Electrochemical Engineering (Université Paul Sabatier, Toulouse, France) for the electrodeposition of tantalum in molten salts and the preparation and characterization of iridium-based industrial electrodes for oxygen evolution in acidic media (sponsored by Electricité de France).

1992-1993 Design engineer, Institute of Marine Biogeochemistry (CNRS & École Normale Supérieure, Paris, France) for the environmental monitoring of heavy-metal pollution by electroanalytical techniques and by alpha spectrometry.

1990-1992 Research scientist, Laboratory of Electrochemistry (Université Pierre & Marie Curie, Paris, France) for the development of a beta nuclear scintillation detector used for electrochemical experiments involving radiolabelled compounds.

"The true university of these days is a collection of books"

Thomas Carlyle

Preface

This monograph is primarily intended to serve as a concise review of the industrial utilization of sulfuric acid and the plethora of sulfation techniques used extensively in the mineral, chemical, and metallurgical industries across the world for more than two centuries.

The information has been presented in such a form that industrial chemists, chemical engineers, and other practicing engineers, scientists, professors, and technologists will have access to relevant scientific and technical information supported by key data gathered from several disseminated sources along with a brief description of each major industrial processes, and novel sulfation technologies that might be implemented commercially in the near future.

I hope this monograph will be of value also to men and women engaged in other branches of chemistry and metallurgy that want to understand these techniques outside their field of expertise.

Finally, the monograph may be of interest to persons in the chemical and metal industries occupying nontechnical positions such as executives, patent attorneys, traders, purchasing agents, salesmen and women, to whom a general knowledge of the technical aspect of their business would be helpful.

Montréal, Québec, Canada   François Cardarelli, December 2022

Dedication

To my former colleague and friend the late Dr. Michel GUÉGUIN who devoted his entire career to invent, patent, implement, and improve the sulfation and chlorination techniques used extensively across the titanium dioxide pigment industry and also to all the men and women who as chemists, engineers, scientists, and technologists from the mineral, chemical, and metallurgical industries devoted their career to invent, devise, implement, and improve industrial sulfation techniques.

Acknowledgements

I want to express my deepest thanks to my late mother Claudine, my father Antonio who together with my late uncle Consalvo supported me in the early 1980s in establishing a basic mineralogical, chemical, and metallurgical laboratory, and scientific library. It is during this period that I was first acquainted with the sulfuric acid digestion, the high pressure sulfuric acid leaching inside an autoclave, and the sulfation roasting of several minerals and ores originating from France (Ardèche) and Italy (Dolomites).

Table of Contents

List of Tables

List of Figures

1 Introduction

Since John Roebuck an Englishman from Birmingham invented and implemented in 1746 the first industrial production of concentrated sulfuric acid by the ***Lead Chamber Process*** [1, 2], nowadays sulfuric acid is exclusively produced by the ***Contact Process***, and has become the largest tonnage chemical produced worldwide [3] with an estimated 231 million tonnes (Mt) in 2012 with China (74 Mt), USA (37 Mt), India (16 Mt), Russia (14 Mt), and Morocco (7 Mt) as leading producing countries [4]. Actually, its annual consumption is regarded together with that of other large tonnage chemicals such as chlorine and titanium dioxide as key economic indicator.

The major utilization of sulfuric acid according the type of industry is roughly as follows: fertilizers (60%), chemicals (25%), titanium dioxide pigment (5%), petroleum refining (5%), textile (3%), and iron and steel surface treatment and finishing (2%).

Despite, the industrial manufacture of sulfuric acid is not the purpose of this monograph, we encourage the reader on consulting old treatises on the matter [5], [6] with still some relevant technical details not reported elsewhere and more recent textbooks [7, 8] including the latest technological developments and improvements along with a comprehensive picture of the current market [9].

---

[1] LUNGE, G.; and NAVILLE, J. (1879) *Traité de la grande industrie chimique, Tome I. Acide sulfurique et oléum.* Masson & Cie, Paris, France.

[2] JONES, E.M. (1950) Chamber Process Manufacture of Sulfuric Acid. *Industrial and Engineering Chemistry*, **42**(11)2208–2210.

[3] GREENWOOD, N.N.; and EARNSHAW, A. (1997) *Chemistry of the Elements, Second Edition.* Butterworth-Heinemann, Oxford, UK.

[4] ANONYMOUS (2013) *Sulfuric Acid: 2013 World Market Outlook and Forecast up to 2017.* Merchant Research & Consulting Ltd., UK.

[5] FRIEND NEWTON, J. (ed.)(1931) *A Text-Book of Inorganic Chemistry, Volume VII, Part II: Sulphur, Selenium, and Tellurium.* Charles Griffin & Company Ltd., London, UK.

[6] FAIRLE, A.M. (1936) *Sulfuric Acid Manufacture.* - Reinhold Publishing Corp. New York, NY.

[7] DAVENPORT, W.G.; and KING, M.J. (2006) *Sulfuric Acid Manufacture, Analysis, Control and Optimization.* Elsevier Science, Amsterdam, The Netherlands.

[8] ASHAR, N.G.; and GOLWALKAR, K.R. (2013) *A Practical Guide to the Manufacture of Sulfuric Acid, Oleums, and Sulfonating Agents.* Springer, Cham, Switzerland.

[9] LOUIE, D.K. (2008) *Handbook of Sulfuric Acid Manufacturing, Second Edition.* DKL Engineering, Inc., Mississauga, ON, Canada.

The sulfuric acid digestion, the sulfuric acid baking, and the sulfation roasting of minerals and metallic ores and concentrates are used in the mineral and chemical processing industries and in extractive metallurgy for more than one hundred years. The peculiar physical and chemical properties of concentrated sulfuric acid along with its wide commercial availability and cost affordability are the main reasons for the major industrial development and implementation of these sulfation methods worldwide.

After reviewing the peculiar physical and chemical properties of concentrated sulfuric acid as a powerful ionizing solvent and a strong sulfating agent, we will describe the suitable corrosion materials, followed by industrial applications of the sulfation categorized by commodities, the latest recent developments in metallurgy with promising new processes, and finally the peculiar prototype and testing methods used.

2 Physical and Chemical Properties of Pure Sulfuric Acid

Prior to describe the various sulfation methods used industrially, it is important to summarize the peculiar physical and chemical properties of concentrated sulfuric acid that lead to the wide utilization of this cost affordable and widely available as a powerful sulfating agent and solvent.

2.1 Chemical Properties

Sulfuric acid, with chemical formula H_2SO_4, and CAS No. [7664-93-9], is a dense (1831 kg/m^3) and viscous liquid (24 mPa.s) with a syrupy consistency with an elevate permittivity (i.e., dielectric constant) of 100 which is almost thirty percent greater than that of pure water that makes it a powerful ionizing solvent for many inorganic and organic compounds.

Moreover, its low surface tension (51 mN/m) which is conversely thirty percent lower than that of pure water ensures a much better wetting of particulates and solids during sulfation [10] when using highly concentrated acid as it increases for aqueous solution due to the water content [11].

Its high boiling point (290°C) allows processing materials at much higher temperatures than when using aqueous solutions of sulfuric acid thus ensuring faster reaction kinetics. In addition, due to its low specific heat capacity (1,399 $J.kg^{-1}K^{-1}$) which is sixty seven percent lower than that of pure water, it necessitates a lower amount of heat to reach a proper operating temperature.

[10] GILLESPIE, B.E.; SMITH, M.J.; and WYATT P.A.H. (1969) Surface tension measurements in solvent sulphuric acid. *Journal of the Chemical Society A: Inorganic, Physical, Theoretical* 0 304-306.

[11] SUGGITT, R.M.; AZIZ, P.M.; and WETMORE, F.E.W. (1949) The surface tension of aqueous sulfuric acid solutions at 25°C. *Journal of the American Chemical Society*, **71**(2)676–678.

Some selected physical properties of pure sulfuric acid (i.e., 100 wt.% H_2SO_4) are listed in Table 1 [12].

Table 1 – Physical properties of concentrated sulfuric

Physical properties (*)	SI value
Mass density (ρ/kg.m^{-3})	1,830.45
Dynamic viscosity (η/mPa.s)	23.55
Surface tension (σ/mN.m^{-1})(30°C)	50.6
Thermal conductivity (k/W.m^{-1}.K^{-1})	0.36
Specific heat capacity (c_p/J.kg^{-1}.K^{-1})	1,399
Electrical conductivity (κ/mS.cm^{-1}) (21°C)	11.8
Permittivity (ε_r)	100
Freezing (melting) point (T_m/K)	283.52 (10.37°C)
Boiling (vaporization) point (T_b/K)	563.15 (290°C)
Refractive index (n_D^{20})	1.4290
Raoult's freezing point depression (/K.mol^{-1}.kg)	6.12
(*) physical properties measured at 20°C unless noted otherwise	

2.2   Dissociation and Ionization

The high electrical conductivity of sulfuric acid is also remarkable when compared to other inorganic solvents. Actually, concentrated sulfuric acid dissociates into the $H_3SO_4^+$ cations and bisulfate anions HSO_4^- according to the following auto-protolysis chemical reaction equation:

$$H_2SO_4 + H_2SO_4 = H_3SO_4^+ + HSO_4^-$$

The above auto-protolysis reaction exhibits a very high auto-protolysis constant especially when compared to that of pure water [$K_{water} = (H_3O^+)(OH^-) = 10^{-14}$]:

$$K_{Auto} = (H_3SO_4^+) \cdot (HSO_4^-) = 2.7 \times 10^{-4}$$

The great ionic mobility of these two ions is the reason of such high electrical conductivity.

---

[12] DEAN, J.A. (1973) *Lange's Handbook of Chemistry, Eleventh Edition.* McGraw-Hill Book Company Inc., New York, NY.

2.3   Hammett Acidity Function

According to the ***Brønsted's theory*** when sulfuric acid is the solvent an acid (HA) is a compounds that gives a proton to yield the cation $H_3SO_4^+$ while a base (B) is a compound that gain a proton and yield the bisulfate anion HSO_4^-.

$$HA + H_2SO_4 = H_3SO_4^+ + A^-$$

$$B + H_2SO_4 = BH^+ + HSO_4^-$$

The dissociation constant of an acid denoted K_{BH+} which is used to measure the acidity of weak acids in aqueous solutions is not applicable to non-aqueous solvents such as concentrated sulfuric acid.

Therefore, a more general measure of acidity valid for any acid is the ***Hammett Acidity Function***, denoted by the symbol Ho [13]. The Hammett acidity function is extends the pH concept for use in non-aqueous solvent and is valid for any acid.

$$Ho = \text{colog}_{10}(a_{H+})$$

Actually, in aqueous solution both quantities are identical. However, the activity of the proton in the Hammett acidity function is far more than a generalization of the pH concept from Brønsted.

Thus the measure of the acidity of a compound in concentrated sulfuric acid [14] corresponds to its ability to protonate a weak basc indicator, denoted B, according to the following reaction:

$$B + H^+ = BH^+$$

This reaction yields the conjugated strong acid BH^+ which ionization is controlled by the activity of protons a_{H+} according to the reverse reaction:

$$BH^+ = B + H^+$$

With the dissociation constant defined as follows:

$$K_{BH+} = (H^+)(B)/(BH^+)$$

---

[13] HAMMETT, L.P. (1928) A theory of acidity. *Journal of the American Chemical Society* **50**(10)2666-2673.

[14] HAMMETT, L.P.; and DEYRUP, A.J. (1932) A series of simple basic indicators. Part I. The acidity functions of mixtures of sulfuric and perchloric acids with water. *Journal of the American Chemical Society* **54**(7)2721-2739.

Rearranging the above equation and using the negative Brigg's logarithm, we obtain the following:

$$H_0 = pK_{BH^+} + \log_{10}(B)/(BH^+)$$

From the above equation similar to the well know ***Henderson-Hasselbach*** equation used for acid-base equilibria in aqueous media, it is apparent that H_0 represents the ability of an acid to donate a hydrogen ion, as measured in terms of its ability to shift the equilibrium between B and BH^+ towards BH^+. Thus more negative values of H_0 correspond to stronger Brønsted acids.

In practice, the value of H_0 is determined experimentally by measuring the ratio of the activities of (BH^+) to (B) by UV-visible, infrared (IR), and nuclear magnetic resonance (NMR) spectroscopic techniques, using weakly basic aromatics with strong electron acceptor groups such as 2,4,6-trinitroaniline, 2,4,6-nitrotoluene, 1,3,5-trinitrobenzene, and trifluoromethylbenzene as the base indicator [15]. Selected Hammett acidity functions for sulfuric acid and other inorganic acids are reported in the Table 2.

Table 2 – Hammett acidity function for sulfuric acid and other inorganic acids

Acid	Chemical formula	Hammett acidity function
Sulfuric acid	H_2SO_4	-12.0
Perchloric acid	$HClO_4$	-13.0
Pyrosulfuric acid	$H_2S_2O_7$	-15.0
Hydrofluoric acid	HF	-15.0
Triflouromethanesulfonic acid	CF_3SO_3H	-14.6
Fluorosulfonic acid	FSO_3H	-15.6

From the above table, we can see that pure sulfuric acid exhibits a Hammett acidity function of -12. Thus arbitrarily, it has been decided that acids with a lower Hammett acidity function than that of pure sulfuric acid ($H_0 < -12$) are called super acids [16].

Thus water for instance, is considered as a base in concentrated sulfuric acid according to the following reaction scheme:

$$H_2O + H_2SO_4 = H_3O^+ + HSO_4^-$$

---

[15] LIANG, J.J.-N. (1976) *The Hammett Acidity Function for Hydrofluoric Acid and some related Superacid Systems.* PhD Dissertation Thesis, McMaster University, Hamilton, ON, Canada.

[16] OLAH, G.A.; SURYA PRAKASH, G.K.; MOLNÁR, A.; and SOMMER, J. (2009) *Superacid Chemistry, Second Edition.* John Wiley & Sons, Inc., Hoboken, NJ.

As a matter of fact, most inorganic compounds considered as acids in aqueous solutions such as ortho-phosphoric acid (H_3PO_4), nitric acid (HNO_3), boric acid (H_3BO_3), sulfurous acid (H_2SO_3), hydrofluoric acid (HF) all behave as bases yielding cations that do no occurs in aqueous solutions according to the following chemical reactions:

$$H_3PO_4 + H_2SO_4 = H_4PO_4^+ + HSO_4^-$$

$$HNO_3 + 2H_2SO_4 = NO_2^+ + H_3O^+ + 2HSO_4^-$$

$$H_3BO_3 + 6H_2SO_4 = 3\ H_3O^+ + B(HSO_4)_4^- + 2HSO_4^-$$

$$H_2SO_3 + H_2SO_4 = SO_2 + H_3O^+ + HSO_4^-$$

$$HF + 2H_2SO_4 = HSO_3F^- + H_3O^+ + HSO_4^-$$

In fact, only a relatively small number of inorganic compounds are considered true acids in concentrated sulfuric acid, these are for instance disulfuric acid ($H_2S_2O_7$), orthoperiodic acid (H_5IO_6), chlorosulfonic (HSO_3Cl) and fluorosulfonic (HSO_3F) acids.

Another important fact, when it comes to the reactivity of concentrated sulfuric acid, is that hot concentrated sulfuric acid reacts with several metal salts to free the volatile acid if the latter is not a reducing or oxidizing agent towards the sulfate anion, this is especially the case with halides such as fluorides, chlorides, perchlorates, nitrates, nitrites, acetates and sulfides liberating hydrogen fluoride (HF), hydrogen chloride (HCl), perchloric acid ($HClO_4$), nitric acid (HNO_3), nitrous acid (HNO_2), acetic acid (CH_3COOH), and hydrogen sulfide (H_2S) gases respectively.

Some key chemical reactions involving the minerals fluorite (fluorspar), halite (rock salt), saltpetre, and pyrite with are reported below some of them still used industrially:

$$CaF_2(s) + H_2SO_4(l) = CaSO_4(s) + 2HF(g)\uparrow$$

$$2NaCl(s) + H_2SO_4(l) = Na_2SO_4(s) + 2HCl(g)\uparrow$$

$$2KNO_3(s) + H_2SO_4(l) = K_2SO_4(s) + 2HNO_3(g)\uparrow$$

$$FeS_2(s) + H_2SO_4(l) = FeSO_4(s) + 2H_2S(g)\uparrow$$

On the other hand, the addition of anhydrous sulfur trioxide gas (SO_3) into concentrated sulfuric acid yields the so-called ***Nordhausen's acid***. In fact, it is not a solution of sulfur trioxide in sulfuric acid but instead a solution of *pyrosulfuric acid* (i.e., disulfuric acid) CAS No. [8014-95-7] according to the following chemical reaction scheme:

$$SO_3(g) + H_2SO_4(l) = H_2S_2O_7(l)$$

In concentrated sulfuric acid, the newly formed pyrosulfuric acid reacts as a Brønsted's acid according to the chemical equilibrium with greater Hammett acidity function reported previously:

$$H_2S_2O_7(l) + H_2SO_4(l) = H_3SO_4^+ + HS_2O_7^-$$

The proportion of pyrosulfuric acid present or eventually formed during sulfation is of paramount importance as in certain novel metallurgical technologies for processing rare earth concentrates and their by-product it can form highly stable solid phases with lanthanide elements and especially thorium pyrosulfate, $Th(S_2O_7)_2$, which is highly insoluble once formed leading to important losses in recovery.

2.4   pH and Acidity of Aqueous Solutions

Regarding the acidity of aqueous solutions of sulfuric acid, the calculation of the pH requires to determine the activity of hydronium ions (H_3O^+) obtained by solving by iterations the following cubic equation with the two dissociation constants $K_{a1} \approx 1000$, and $K_{a2} = 0.01$:

$$[1/(K_{a1} + K_{a2})](H_3O^+)^3 + (1/K_{a2})(H_3O^+)^2 + (1 - C_A/K_{a2})(H_3O^+) - 2\,C_A = 0$$

Table 3 – pH of aqueous solutions of sulfuric acid

Molarity	Normality	pH(calc.)	pH(actual)
5.00 M	10.0 N	-0.70	-0.92
4.00 M	8.0 N	-0.60	-0.80
3.00 M	6.0 N	-0.48	-0.56
2.50 M	5.0 N	-0.40	-0.40
2.25 M	4.5 N	-0.35	-0.29
2.00 M	4.0 N	-0.30	-0.23
1.50 M	3.0 N	-0.18	-0.10
1.00 M	2.0 N	0.00	0.12
0.75 M	1.5 N	0.12	0.28
0.50 M	1.0 N	0.30	0.51
0.25 M	0.5 N	0.60	0.72
0.10 M	0.2 N	0.99	1.05

The plot of the pH of aqueous solution of sulfuric acid vs. the molarity is depicted in Figure 1.

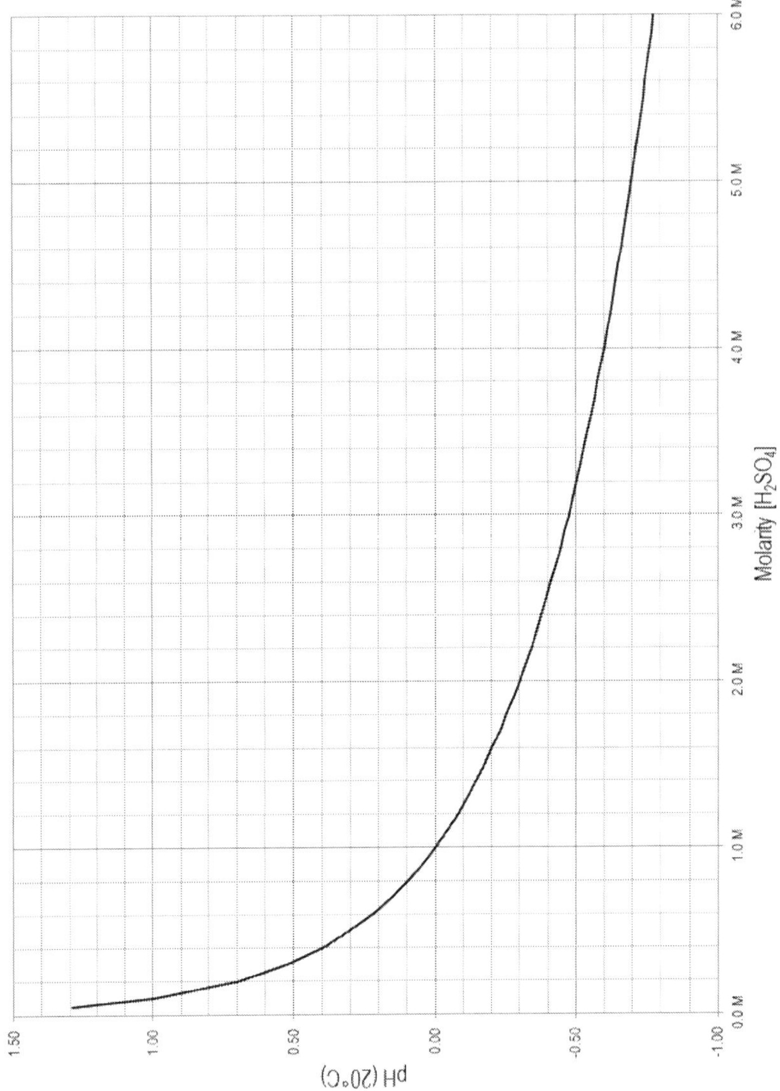

Figure 1 – pH of aqueous solutions of sulfuric acid vs. molarity

When dissolved into water, the degree of dissociation of aqueous solutions of sulfuric acid, that is, the predominance of sulfate (SO_4^{2-}), and bisulfate anions (HSO_4^-) together with non-dissociated sulfuric acid (H_2SO_4) is a function of the molarity of sulfuric acid and it is reported in Table 4 [17] from experimental data obtained by Young and Blatz [18] using Raman scattering spectroscopy.

Table 4 – Predominance of sulfate, bisulfate anions, and sulfuric acid vs. molarity

Sulfuric acid molarity (M)	Mass percentage (wt.% H_2SO_4)	SO_4^{2-}	HSO_4^-	H_2SO_4
0.05	0.49%	100.0%	nil	nil
1	8.89%	11.7%	85.1%	3.2%
2	16.34%	6.4%	88.3%	5.3%
3	22.65%	4.3%	88.3%	7.4%
4	28.10%	3.2%	87.2%	9.6%
5	32.86%	3.2%	85.1%	11.7%
7.5	42.27%	2.2%	77.7%	20.1%
10	49.47%	2.1%	67.0%	30.9%
15	59.48%	nil	18.1%	81.9%
18	63.72%	nil	2.1%	97.9%

2.5   Redox Properties

The oxidation and reduction properties of aqueous solution of sulfuric acid can be illustrated by the Pourbaix (Eh-pH) diagram [19] describing the sulfur system (Figure 2):

Dilute acidic solutions of sulfuric acid exhibit the electroactive domain of pure water (1.229 V at 298.15K) because of the wider decomposition limits (i.e., electrochemical span) for the sulfate anions.

---

[17] HARNED. H.S.; and OWEN, B.B. (1958) *The Physical-Chemistry of Electrolytic Solutions, Third Edition.* American Chemical Society Monograph Series, Reinhold, New York, NY, pp. 592.

[18] YOUNG, T.F.; and BLATZ, L.A. (1949) The variation of the properties of electrolytic solutions with degrees of dissociation. *Chemical Review*, **44**(1)93-115.

[19] POURBAIX, M. (1974) – *Atlas of Electrochemical Equilibria in Aqueous Solutions.* National Association of Corrosion Engineers (NACE) International Houston, TX, and CEBELCOR, Brussels, Belgium, pp. 551.

Thus the lower potential limit of such solutions is imposed by the cathodic decomposition of water due to the hydrogen gas evolution reaction (HER) at 298.15K and 101.325 kPa:

$$2H^+ + 2e^- = H_2(g)\uparrow \qquad \text{with} \qquad E(V/SHE) = -0.059pH$$

The upper potential limit is imposed by the anodic decomposition of water due to the oxygen gas evolution reaction (OER) at 298.15K and 101.325 kPa:

$$2H_2O = O_2(g)\uparrow + 4H^+ + 2e- \qquad \text{with} \qquad E(V/SHE) \;=\; 1.229 - 0.059pH$$

However, for strongly acidic solutions of sulfuric acid with a mass percentage above 50 wt.% H_2SO_4, the sulfate anions become electroactive.

For hot sulfuric acid solution with concentration above 50 wt.% H_2SO_4, the sulfate anions are reduced at 140°C by sparging hydrogen or carbon monoxide gases yielding sulfur dioxide gas (SO_2) according to the half-reaction:

$$SO_4^{2-} + 4H^+ + 2e^- = SO_2(g) + 2H_2O$$

With $E(V/SHE) = 0.170 + 0.030\,\log_{10}\{[SO_4^{2-}]/p_{SO2}\} - 0.059\ pH$

This reaction occurs even at a lower temperature when concentrated sulfuric in electrolyzed inside the cathode compartment of a two-compartment electrolyzer acid due to the high reactivity of nascent hydrogen gas evolving at a cathode surface.

In some instance, this reduction proceeds further to yield elemental sulfur (S_8) or even hydrogen sulfide (H_2S) when hot concentrated sulfuric acid reacts with hydroiodic acid (HI), zinc (Zn) or iron (Fe) metal powders:

$$SO_4^{2-} + 10H^+ + 2e^- = H_2S(g) + 5H_2O$$

Hot concentrated sulfuric acid is also reduced rapidly by carbon (e.g., coal) and finely divided sulfur with the strong and copious gas evolution of carbon dioxide, sulfur dioxide, and steam according to the following reactions:

$$C(s) + 2\ H_2SO_4(l) = CO_2(g) + 2SO_2(g) + 2H_2O(g)$$

$$S(s) + 2H_2SO_4(l) = 3SO_2(g) + 2H_2O(g)$$

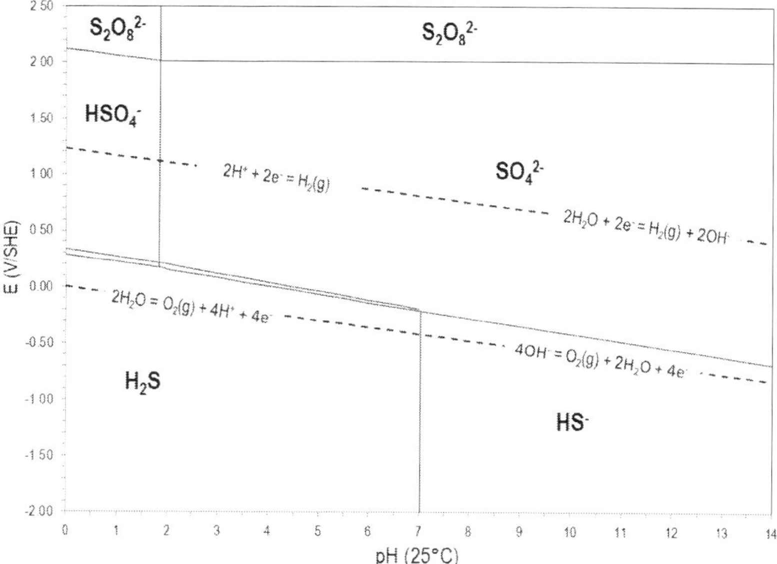

Figure 2 – Simplified Pourbaix (E-pH) diagram of sulfur

On the other hand, under strongly oxidizing conditions, the sulfuric acid is oxidized into peroxodisulfuric acid or ***Marshall's acid*** with the chemical formula $H_2S_2O_8$, CAS No. [13445-49-3] according to the following half-reaction [20]:

$$2SO_4^{2-} - S_2O_8^{2-} + 2e^- \qquad E(V/SHE) = 2.010 - 0.030 \log_{10}\{[S_2O_8^{2-}]/[2SO_4^{2-}]\}$$

The above oxidation reaction takes place essentially inside the anode compartment of an electrolyzer equipped with smooth platinum metal or mixed metal oxides (MMO) anodes of the type $Ta/IrO_2-Ta_2O_5$. However, because of the fast decomposition and poor thermal stability of the peroxodisulfuric acid at temperature above 40°C[21], to obtain current efficiencies above 65%, the concentrated sulfuric acid must be electrolyzed at low temperatures well below 10°C and under elevate anode current density (+ 3000 A/m²)[22].

[20] HOLLEMAN, A.F.; and WIBERG, N. (2007) *Lehrbuch des Anorganishen Chemie, 102 Auflage.* Walter De Gruyter, Berlin, pp. 600-601.

[21] WAHL, H. (1973) *Elements de Chimie Minerale.* Collection du CNAM, Masson & Cie, Paris, France. pp. 29-30.

[22] MANTELL, C.L. (1950) *Industrial Electrochemistry, Third Edition.* Chemical Engineering Series, McGraw-Hill Book Company Inc., New York, NY, pp. 110-111.

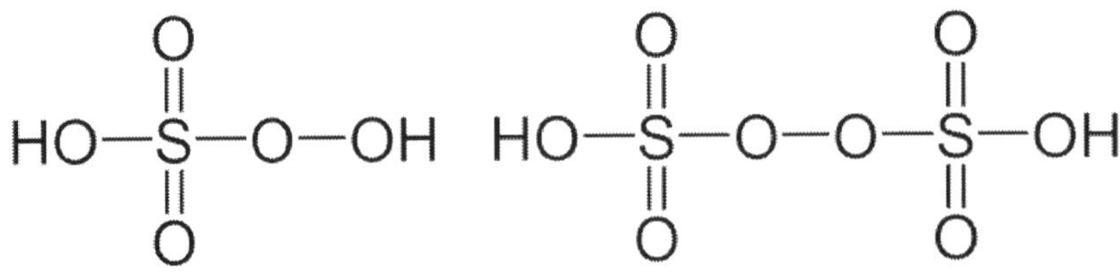

Figure 3 – Structure of Caro's acid   Figure 4 – Structure of Marshall's acid

On the other hand, mixing hydrogen peroxide, H_2O_2, with concentrated sulfuric acid or absorbing sulfur trioxide gas directly into concentrated hydrogen peroxide yields the peroxomonosulfuric acid or **Caro's acid** with the chemical formula H_2SO_5, CAS No. [7722-86-3].

In fact, by contrast to the H_2O-SO_3 system used to describe sulfuric acid solutions, the Caro's acid is more accurately described using the binary system H_2O_2-SO_3.

The production of Caro's acid occurs according to the following fast and strongly exothermic chemical reaction:

$$H_2SO_4 + H_2O_2 \rightarrow H_2SO_5 + H_2O$$

Actually, when the two reagents are mixed at ambient temperature, the resulting mixture containing Caro's Acid reaches immediately a temperature in the 110-120°C (230-248°F) range.

Caro's acid which is a highly reactive compound and powerful oxidizing agent has been used during the oxidizing sulfuric acid leaching of minerals and concentrates (e.g., uranium, rare earths) and as bleaching agent in the pulp and paper industry.

Since the late 1990s, it started being used also as in the detoxification of wastewaters containing cyanides, a technique adopted in several plants of gold extraction in France, Peru and the United States.

However, Caro's acid decomposes easily [23] especially in presence of traces of deleterious impurities (e.g., Cu, Ni, Pt) into sulfuric acid and nascent oxygen gas according to:

$$2H_2SO_5 \rightarrow 2H_2SO_4 + O_2(g)$$

---

[23] BALL, D.L.; and EDWARDS, J.O. (1956) The kinetics and mechanism of the decomposition of Caro's acid. Part I. *Journal of the American Chemical Society*, **78**(6)1125–1129.

Thus the chemical must be utilized immediately after its generation for the maximum efficiency.

The presence of the two peracids can be easily distinguished from hydrogen peroxide as they do not decolorize an aqueous solution of potassium permanganate ($KMnO_4$).

Marshall's acid is not as a strong oxidant compared to Caro's acid. Actually, the free acid prepared from a boiling mixture of strong sulfuric acid and ammonium persulfate $(NH_4)_2S_2O_8$ needs traces of silver (I) cations to be oxidized first into Ag(II) capable to oxidize readily Mn(II) into Mn(VI), Cr(III) into Cr(VI), and V(IV) into V(V). Without the traces of silver cations the mixture is unable to perform this task. This technique was used extensively in the quantitative chemical analysis by visible spectrophotometry of manganese, chromium, and vanadium in cast irons and steels [24].

2.6 Solvating Properties and Metal Sulfates Solubilities

As mentioned earlier, because of its elevate dielectric constant, concentrated sulfuric acid possesses strong ionizing capabilities that ensure the solubilisation of several inorganic salts especially metal sulfates salts including some considered almost insoluble in dilute aqueous solutions of the acid (e.g., $BaSO_4$) while other reputed fully soluble in aqueous solution of sulfuric acid become almost insoluble (e.g., $MgSO_4$, $FeSO_4$, $NiSO_4$) in that solvent.

The solubility of selected metal sulfates in pure sulfuric acid were extensively measured experimentally by James Kendall, Mary Louise Landon, William Davidson, and Howard Adler [25, 26, 27].

When expressed in mole and mass percent respectively the solubility extends over three decades. For instance some selected values adapted from [28] and other sources ordered according to mass percentages are reported in Table 5.

From the above data, we can distinguish clearly three categories of anhydrous metal sulfates:

---

[24] VOGEL, A. I. (1961) *Textbook of Quantitative Inorganic Analysis, Third Edition.* Longman, London, UK, pp. 817-820.

[25] KENDALL, J. and LANDON, M.L., (1920) The formation of addition compounds between 100% sulfuric acid and the neutral sulfates of the alkali metals. *Journal of the American Chemical Society* **42**(11)2131-2142.

[26] KENDALL, J.; and DAVIDSON, A.W (1921) Compound formation and solubility in systems of the type sulfuric acid: metal sulfate. *Journal of the American Chemical Society* **43**(5)979–990.

[27] KENDALL, J.; DAVIDSON, A.W., J.; and ADLER, H. (1921) The prediction of solubility in polar solutions. *Journal of the American Chemical Society* **43**(7)1481-1502.

[28] NYEIN, T.T. (1962) *Chemistry of Titanium Compounds in Sulfuric Acid.* PhD Thesis, University of London, UK, page 20.

---

(1)   The first category regroups anhydrous metal sulfates with a strong propensity to dissolve in concentrated sulfuric acid with their solubility ranging from 7 mass percent for calcium sulfate up to 50 mass percent for titanium (III) sulfate. Beside these elevate solubilities, it must be noted that some of these metal sulfates especially barium sulfate and in a lesser extend calcium sulfate that are particularly insoluble in aqueous solutions of sulfuric acid shows rather high solubilities. As a general rule, most of these soluble metal sulfates can be recovered from the sulfuric acid as bisulfates ($MHSO_4$) but most of them form solid phase in equilibrium that are in fact adduct products such as $MHSO_4.H_2SO_4$ for alkali-metals with $M = Li^+$, Na^+, K^+, and $M(HSO_4)_2.2H_2SO_4$ for alkaline earth metals with $M = Ca^{2+}$, Mg^{2+}, Sr^{2+}, and Ba^{2+}. At this point, it is important to mention that the elevate solubilities of titanium (III) sulfate along with that of lithium sulfate are one of the reasons for explaining the industrial success of the sulfation processes used for recovering titanium from ilmenite and titanium slags and lithium from spodumene concentrates.

(2)   The second category regroups anhydrous metal sulfates with solubilities ranging from 0.1 mass percent up to 1.0 mass percent. Among them anhydrous sulfates of Mg, Fe an Cu, Zn, and Ti(IV).

(3)   The third category regroups anhydrous metal sulfates with solubilities below 0.05 mass percent that are considered totally insoluble. These are mostly anhydrous sulfates of the chemical elements of group IIIA(13) such as Al, Ga, Tl, and also Sc from group IIIB(3).

Table 5 – Solubility of metal sulfates in pure concentrated H_2SO_4 at 25°C

Metal sulfates	Relative molar mass (M_R)	Mole percent (mol.%)	Mass percent (wt.%)	kg of salt per 100 kg of acid
$Ti_2(SO_4)_3$	383.925	20.51	50.2	101.00
Ag_2SO_4	96.064	9.11	24.2	31.86
$BaSO_4$	96.064	8.85	18.8	23.10
Li_2SO_4	109.946	14.28	15.7	18.67
K_2SO_4	174.260	9.24	15.3	18.09
Na_2SO_4	142.043	5.28	7.5	8.11
$CaSO_4$	136.142	5.16	7.0	7.53
$HgSO_4$	96.064	0.78	0.80	0.806
$TiOSO_4$	159.930	0.51	0.82	0.830
$ZnSO_4$	161.454	0.17	0.28	0.281
$FeSO_4$	151.909	0.17	0.26	0.261
$MgSO_4$	120.369	0.18	0.22	0.220
$CuSO_4$	159.610	0.08	0.13	0.130
$(VO_2)_2(SO_4)$	261.944	0.05	0.13	0.130
$PbSO_4$	96.064	0.12	0.12	0.120
$Tl_2(SO_4)_3$	288.191	0.01	0.07	0.071
$VOSO_4$	163.005	0.02	0.04	0.035
$Ga_2(SO_4)_3$	427.637	0.01	0.04	0.044
$Sc_2(SO_4)_3$	378.109	0.01	0.04	0.039
$Al_2(SO_4)_3$	334.170	0.01	0.03	0.034
Hg_2SO_4	96.064	0.02	0.02	0.020
$NiSO_4$	159.610	0.005	0.01	0.008

On the other hand, it is interesting to mention that oxides of refractory metals of group VB (5) such as those of niobium and tantalum which are extremely resistant to attack by many chemicals except hydrofluoric acid and strong caustics become relatively soluble in hot concentrated sulfuric acid, oleum and molten bisulfates, and pyrosulfates [29].

For instance, niobium and tantalum pentoxides dissolve into concentrated sulfuric acid as basic sulfates with the chemical formula $Nb_2O_3(SO_4)_2$ and $Ta_2O_3(SO_4)_2$ respectively. Because these conditions are often encountered during

[29] GOROSHENKO, Ya.G. (1965) *Khimiya niobiya i tantala*: Naukova Dumka, Kiev, Ukraine.

sulfation baking and roasting of tantalum and niobium such as in the processes described in Section 9.10 they are worth mentioning.

Similarly, vanadium (V) oxide (V_2O_5) is slightly soluble hot concentrated sulfuric acid and forms dark red solutions of various adduct products such as: $V_2O_5 \cdot SO_3$, $V_2O_5 \cdot 2SO_3$, $V_2O_5 \cdot 3SO_3$, and $V_2O_5 \cdot 4SO_3$ [30]. The solubility of free V_2O_5 in concentrated sulfuric acid is reported 0.054 wt.% (535 mg/kg)[31] when measured at 30°C.

[30] COLLECTIVE (1967) *Gmelin Handbuch der Anorganishen Chemie. Vanadium Teil B – Lieferung 1. System-Nummer, 8 Auflage.* Verlag Chemie GmbH, Weinheim, Germany, pp. 302-303.

[31] LANFORD, O.E.; and KIEHL, S.J. (1940) A study of heterogeneous equilibria in aqueous solutions of the sulfates of pentavalent vanadium at 30°C. *Journal of the American Chemical Society* (JACS) **62**(7)1660-1665.

3 Binary Phase Diagram

3.1   Binary Phase Diagram Water-Sulfuric Acid

In industrial practice pure 100 wt.% sulfuric acid is seldom used and most of the time the sulfuric acid utilized consists of mixtures of sulfuric acid and water. Actually, the strong hygroscopic behavior of sulfuric acid for water implies the formation of definite crystalline adduct compounds with a precise stoichiometry called sulfuric acid hydrates.

Several researchers have investigated the phase equilibria and the recording of the freezing point of the mixtures of water-sulfur dioxide [32]. The simplified binary phase diagram for the system water-sulfuric acid (H_2O-H_2SO_4) is depicted in Figure 5 [33].

A more recent version for the water-sulfur trioxide binary phase diagram accounting for new hydrates was published by Zeleznik [34]. In the latter publication, the comprehensive phase diagram reveals that beside water (H_2O) and sulfur trioxide (SO_3) as the two components extrema, twelve crystalline phases and eutectics were identified so far and numbered with Roman numerals from I to XII according to their increasing content of sulfur trioxide.

[32] KUNZLER, J.E.; and GIAUQUE, W.F. (1952) The freezing point curves of concentrated aqueous sulfuric acid. *Journal of the American Chemical Society* **74**(21)5271-5274.

[33] GABLE, C.M.; BETZ, H.F.; and MARON, S.H. (1950) Phase equilibria of the system sulfur trioxide-water. *Journal of the American Chemical Society*, **72**(4)1445-1448.

[34] ZELEZNIK, F.J. (1991) Thermodynamic properties of the aqueous sulfuric acid system to 350K. *Journal of Physical and Chemical Reference Data*, **20**, 1157-1199.

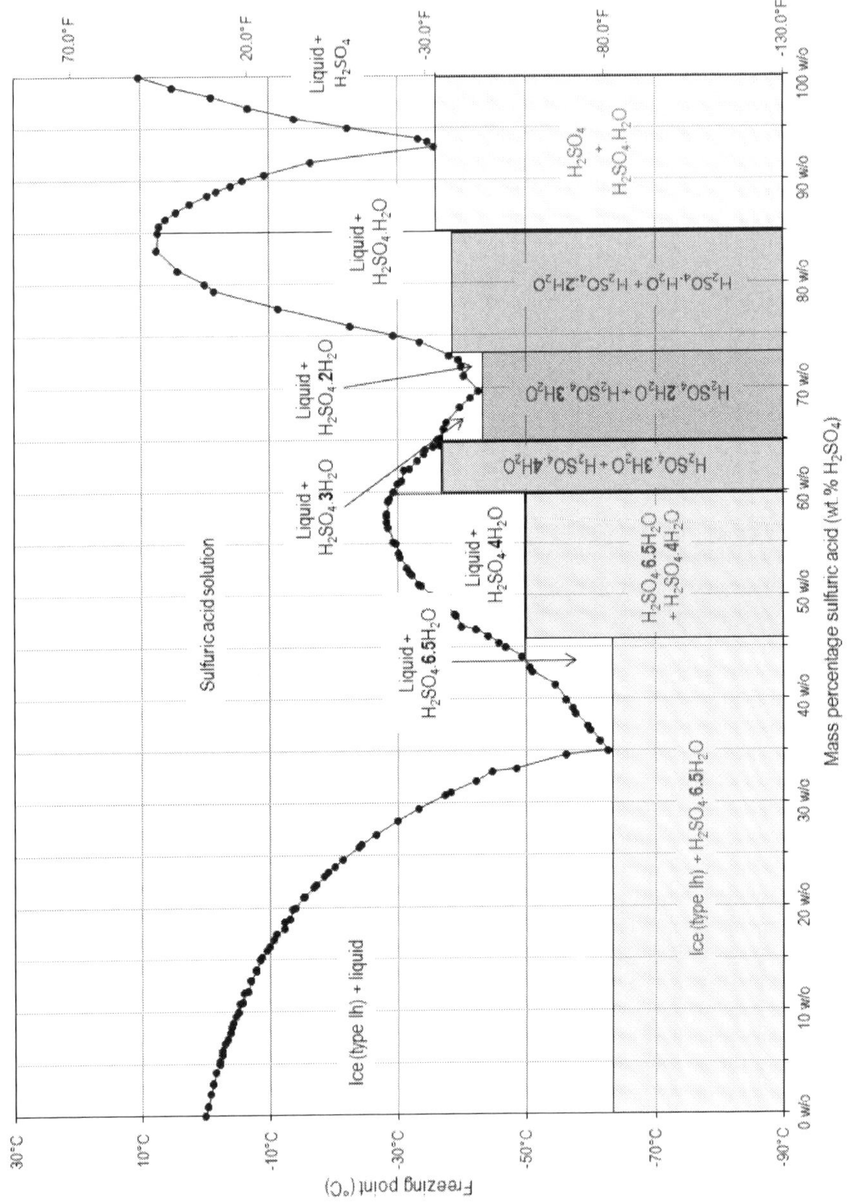

Figure 5 – Binary phase diagram of H₂O-H₂SO₄ system

The following compounds $H_2S_2O_7$, H_2SO_4, $H_2SO_4 \cdot H_2O$, $H_2SO_4 \cdot 2H_2O$, $H_2SO_4 \cdot 3H_2O$, $H_2SO_4 \cdot 4H_2O$, and $H_2SO_4 \cdot 6.5H_2O$, with their salient properties are reported in Table 6.

Table 6 – Phases identified in the H_2O-SO_3 phase diagram

Phase	Compound	Relative molar mass ($^{12}C = 12$)	Molar enthalpy of formation ($\Delta H°/kJ.mol^{-1}$) (*)	Molar entropy of formation ($\Delta S°/kJ.mol^{-1}.K^{-1}$) (*)	Mass percent (wt.% H_2SO_4)	Mass percent (wt.% SO_3)	Density 20°C ($\rho_A/kg.m^{-3}$)	Freezing point ($fp/°C$)
I	H_2O	18.01	-285.83	69.95	0.00	0.00	998.204	0.00
III	$H_2SO_4.6.5H_2O$	215.18	-2733.3	587.82	45.58	37.21	1,353.0	-52.88
V	$H_2SO_4.4H_2O$	170.14	-2011.2	414.53	57.65	47.06	1,473.1	-28.27
VII	$H_2SO_4.3H_2O$	152.13	-1720.1	345.37	64.47	52.63	1,547.4	-36.39
IX	$H_2SO_4.2H_2O$	134.11	-1427.1	276.36	73.13	59.70	1,647.1	-39.48
XI	$H_2SO_4.H_2O$	116.10	-1127.6	211.51	84.48	68.96	1,773.8	+8.48
XIII	H_2SO_4	98.08	-814.0	156.90	100.0	81.63	1,830.5	+10.35
XV	$H_2S_2O_7$ ($H_2SO_4.SO_3$)	178.14	-1272.4		55.06	89.89	1,935.0	+36.01
XVII	SO_3	80.06	-454.51	70.7	0.00	100.00	1,920.0	+18.02

Notes: (*) Standard conditions of temperature and pressure (STP): T = 298.15K and p = 101.325 kPa [35]

Moreover, the cryohydrate (i.e., eutectics with ice) and seven eutectics between the above phases were also identified and presented in Table 7.

[35] HORNUNG, E.W.; BRACKETT, T.E.; and GIAUQUE, W. F. (1956) The low temperature heat capacity and entropy of sulfuric acid hemihexahydrate. Some observations on sulfuric acid "octahydrate" *Journal of the American Chemical Society* **78**(22)5747–5751.

Table 7 – Eutectics identified in the H₂O-SO₃ phase diagram

Phase	Cryohydrates and eutectics	Mass percent (wt.%) H_2SO_4	Mass percent (wt.%) SO_3	Density 20°C (ρ/kg.m^{-3})	Freezing point (fp/°C)
II	H_2O - $H_2SO_4.6.5H_2O$				
IV	$H_2SO_4.6.5H_2O$ - $H_2SO_4.4H_2O$	37.98	31.0	1,285.3	-75
VI	$H_2SO_4.4H_2O$ - $H_2SO_4.2H_2O$	67.99	55.5	1,587.3	-50
VIII	$H_2SO_4.2H_2O$ - $H_2SO_4.H_2O$	73.13	59.7	1,647.1	-41
X	$H_2SO_4.H_2O$ - H_2SO_4	93.10	76.0	1,828.2	-38
XII	H_2SO_4 - $H_2S_2O_7$	104.13	85.0	1,928.1	-12
XIV	$H_2S_2O_7$ - SO_3	113.44	92.6	1,895.1	+0.7

3.2 Azeotropic Diagram Water-Sulfuric Acid

The *azeotrope* of the water-sulfuric acid system has a constant boiling point because the vapor phase exhibits the same composition as the aqueous solution and no separation of the pure component can be made.

In the particular case of the water-sulfuric acid system at 101.325 kPa, the azeotrope has a fixed composition of 98.479 wt.% H_2SO_4, and a boiling point of 338°C, that is, is higher than that of pure water (100°C) and pure sulfuric acid (290°C). Thus, it is a typical example of a negative azeotrope. Negative azeotropes are also called *maximum boiling mixtures* or *pressure minimum azeotropes*.

3.2.1 Influence of Pressure

The influence of pressure on the constant boiling point of water-sulfuric acid mixtures was accurately measured by Kunzler [36] and some selected values are reported in Table 8. We can see from these experimental data that the azeotrope concentration of sulfuric acid decreases slightly with absolute pressure increase.

---

[36] KUNZLER, J.E. (1953) Absolute sulfuric acid, highly accurate primary standard: constant boiling sulfuric acid and other reference standards *Analytical Chemistry*, **25**(1)93-103.

Table 8 – Water-sulfuric acid azeotrope composition vs. absolute pressure

Absolute pressure (mmHg)	Absolute pressure (kPa)	Mass percent H_2SO_4 (wt.%)
100	13.332	98.790
200	26.664	98.704
300	39.997	98.645
400	53.329	98.597
500	66.661	98.557
600	79.993	98.524
700	93.326	98.495
760	**101.325**	**98.479**
800	106.658	98.469
900	119.990	98.446

3.2.2 Influence of Temperature

On the other hand, Abel [37] collected the temperature and composition of the water-sulfuric acid azeotrope at atmospheric pressure measured by several researchers. These are reported in

Table 9 – Water-sulfuric acid azeotrope composition and temperature (atm.)

Absolute temperature (K)	Temperature (°C)	Mass percent H_2SO_4 (wt.%)
590.15	317	98.54
599.15	326	98.39
603.15	330	98.33
611.15	338	98.30

[37] ABEL, J. (1946) The vapor phase above the system sulfuric acid–water. *Journal Physical Chemistry*, **50**(3)260-283.

4 Sulfuric Acid and Oleums

4.1 Strengths of Sulfuric Acid

In the technical literature and data tables used in the trade, the various concentrations of sulfuric acid or strengths are reported as follows:

(1) Its mass percentage (wt.% H_2SO_4);

(2) Its mass density, ρ_A, in kg/m³ (lb/ft³) measured at 60°F (ca. 15.56°C) in North America or at 15°C or 20°C elsewhere;

(3) Its specific gravity, $S.G.(^{60°F}/_{60°F})$, that is, the dimensionless ratio of its mass density measured at 60°F (ca. 15.56°C) compared to that of pure water at 60°F (i.e., 999.013 kg/m³ at 15.56°C);

$$S.G.(^{60°F}/_{60°F}) = \rho_A{}^{60°F}(kg/m^3)/999.013 = \rho_A{}^{60°F}(lb_m/ft^3)/62.3663$$

(4) Its specific gravity, $d(^{20°}/_{4°})$, that is, the dimensionless ratio of its mass density measured at 20°C compared to that of pure water at its maximum density (i.e., 999.973 kg/m³ at 3.98°C) both defined by the following equations:

$$d(^{20°}/_{4°}) = \rho_A{}^{20°C}(kg/m^3)/999.973 = \rho_A{}^{20°C}(lb_m/ft^3)/62.4263$$

(5) Finally, by using an empirical and historical scale called the ***Baumé scale*** [38] for liquids heavier than water (*American scale*) [39] which is still used among the trade and defined as follows:

$$°Bé(60°F) = 145 - 145/[S.G.(^{60°F}/_{60°F})]$$

---

[38] The two Baumé scales were devised in 1768 by Antoine Baumé, a French pharmacist, to measure the density of various aqueous solutions. The Baumé degrees for heavy liquids originally represented the percent by mass of sodium chloride in water at 60°F (16°C) with 0°Bé (heavy) stated as the density of 10 wt.% NaCl in water while the Baumé degrees for light liquids was stated at 10°Bé (light) set to the density of pure water.

[39] CARDARELLI, F. (2005) *Encyclopaedia of Scientific Units, Weights and Measures. Their SI equivalences and origins.* Springer, London, UK, pp. 712-713.

For the industrial usage, the determination of the density of solutions of sulfuric acid using calibrated hydrometers, liquid pycnometers or ultrasonic velocity probes is a reliable and accurate method for measuring the concentration.

However, despite the density increases and reaches a maximum at 97 wt.% H_2SO_4 above that mass percentage threshold, it decreases again meaning the titer of sulfuric acid with higher concentrations must be measured by other techniques such as electrical conductivity, freezing point depression, gravimetric analysis or by potentiometric acid-base titration. This is the reason why the Baume degrees are not reported above that of the oil of vitriol (O.V.) (see below) to avoid confusion.

4.2   Commercial Grades of Sulfuric Acid

The so-called *virgin acid* in the trade denotes sulfuric acid that has never been used industrially in a process. It is commonly called *commercial grade sulfuric acid*. Two grades of virgin or commercial grade acid are industrially important. They are:

(1)   The 66°Bé acid (93.19 wt.% H_2SO_4) also formerly called *Oil of Vitriol* (O.V.)

(2)   The 98-99% H_2SO_4

Table 10 – Commercial grades of concentrated sulfuric acid

Sulfuric acid grade	Mass percent (wt.% H_2SO_4)	Density 20°C (ρ_A/kg.m^{-3})	Density 15.56°C (ρ_A/kg.m^{-3})	Specific gravity $d_{(20°/4°)}$	Specific gravity $S.G._{(60°F/60°F)}$	Baumé degree (°Bé)	Freezing point (fp/°C)
Fertilizer acid	62.18	1,522.0	1,525.8	1.5220	1.5273	50	-32.78
Glover (Tower) acid	77.67	1,700.5	1,704.9	1.7005	1.7066	60	-11.33
Oil of vitriol (O.V.)	93.19	1,828.5	1,833.6	1.8285	1.8354	66	-33.88
98% sulfuric acid	98.00	1,836.1	1,839.6	1.8361	1.8414	w/o	+3.00
100% sulfuric acid	100.00	1,830.5	1,835.4	1.8305	1.8372	w/o	+10.35

Because the 98-99% sulfuric acid has a high freezing point near 10°C, it is usually shipped in insulated steel containers equipped with steam coils. By contrast, the 66°Bé sulfuric acid has a freezing point of -29°C (-20.2°F) and for that reason it is sometimes called *"winter acid"* in the trade because it can be transported in Northern climate regions of the USA, Canada, Russia, and Scandinavia.

Moreover, traces of deleterious impurities originating from the various type of concentrates and raw materials processed such as arsenic, selenium, mercury, and iron are usually tightly controlled within specifications those values are met by modern plant using both proper removal processes and corrosion resistant materials.

4.3   Pyrosulfuric Acid, Fuming acid and Oleums

On the other hand, the addition of anhydrous sulfur trioxide gas (SO_3) into concentrated sulfuric acid yields the so-called *fuming sulfuric acid* or *Nordhausen's acid* [40] that was historically prepared by dissolving gaseous SO_3 produced from the oxidative roasting of iron sulfides (e.g., pyrite and chalcopyrite)

---

[40] HAWLEY, G.G. (ed.) (1977) *The Condensed Chemical Dictionary, Ninth Edition*. Van Nostrand Reinhold Company, New York, NY, page 624.

or the decomposition of iron sulfate(s) in Nordhausen, Germany into 66°Bé (93.19 wt.% H_2SO_4) sulfuric acid coming out from the lead chambers process.

Nowadays it is produced by the direct contact process and it is called fuming sulfuric acid or simply *oleum* [41]. Thus, in fact, it is not a solution of sulfur trioxide in sulfuric acid but instead a solution of *pyrosulfuric acid* (i.e., disulfuric acid) CAS No. [8014-95-7] according to the reaction scheme:

$$SO_3(g) + H_2SO_4(l) = H_2S_2O_7(l)$$

4.4 Commercial Grades of Oleums

Because pyrosulfuric acid is a solid at room temperature with a melting point of 36.01°C (96.82°F), liquid oleums of various compositions are used instead. Oleums can be described by the empirical chemical formula $H_2SO_4 \cdot xSO_3$.

The stoichiometric coefficient x varies depending of the oleum's strength. However, in practice, the strength of oleums is usually reported as a mass percentage of free SO_3 in the mixture that is:

$$wt.\% \ SO_3(free) = 100 \ xSO_3/(H_2SO_4 + xSO_3)$$

Thus, the simple relation between the stoichiometric coefficient x and the mass percentage of free sulfur trioxide, in wt.% SO_3, using the *relative molar masses* of sulfuric acid (M_{H2SO4} = 98.08) and sulfur trioxide (M_{SO3} = 80.06) respectively, is given by:

$$x = 1/\{(M_{SO3}/M_{H2SO4})[100/(wt.\% \ SO_3(free)) - 1]\}$$

$$x \approx 1/\{0.81627 \ [100/(wt.\% \ SO_3(free)) - 1]\}$$

Example: an oleum containing 20 wt.% SO_3 is equivalent to the empirical chemical for formula $H_2SO_4 \cdot 0.306SO_3$. Conversely, in the case of pure pyrosulfuric acid (*senso stricto*) with $x = 1$, the theoretical mass percentage of free sulfur trioxide is thus 44.94 wt.% SO_3.

Additionally, oleums can also be described by the empirical chemical formula $(1 + x)SO_3 \cdot H_2O$. In this case, the total mass percentage of sulfuric acid is given by:

$$wt.\% \ SO_3(total) = 100 \ (1 + x)SO_3/[H_2O + (1 + x)SO_3]$$

---

[41] DUECKER, W.W.; and WEST J.R. (1971) *The Manufacture of Sulfuric Acid*. American Chemical Society (ACS) Monograph Series, Robert E. Kringer Publishing Co., Inc., Huntington, NY.

The relationship between the stoichiometric coefficient x and the total mass percentage of total sulfur trioxide is then given by the following equations:

$$x = (M_{H2O}/M_{SO3})[\text{wt.\% } SO_3(\text{total})/(100 - \text{wt.\% } SO_3(\text{total}))] - 1$$

$$x \approx 0.2258[\text{wt.\% } SO_3(\text{total})/(100 - \text{wt.\% } SO_3(\text{total}))] - 1$$

Example: an oleum containing 40 wt.% SO_3 is thus equivalent to the empirical chemical formula $1.817SO_3 \cdot H_2O$.

Therefore upon dilution with the stoichiometric amount of water to react with the free sulfur trioxide in order to obtain 100% H_2SO_4, the mass ratio of sulfuric acid to the initial mass of oleum is given by:

$$m_{H2SO4}/m_{oleum} = 100 (1 + x) H_2SO_4/[H_2O + (1 + x)SO_3]$$

$$= 100 (1 + x) H_2SO_4/(H_2SO_4 + xSO_3)$$

The above practical equation allows calculating the mass of sulfuric acid per unit mass of oleum expressed as percentage that will be obtained by dilution. Thus this number is always over 100% [42].

Example: an oleum containing 65 wt.% SO_3 is equivalent to the empirical chemical for formula $H_2SO_4 \cdot 2.275SO_3$ or $3.275SO_3 \cdot H_2O$, upon dilution it will yield 3.275 moles of H_2SO_4, that is, a yield of 114.6% sulfuric acid.

The selected chemical composition and physical properties of several oleum grades from various sources [43, 44] are reported in Table 11.

[42] FASULLO, O.T. (1965) *Sulfuric Acid: Use and Handling*. McGraw-Hill Book Company, New York, NY.

[43] CASTELL-EVANS, J. (1911) *Physico-chemical Tables. Volume II Analytical Chemistry*. Charles Griffin & Company Limited, Strand, UK, pp. 843-862

[44] GAMBERINI, G. (1960) *Il Memorandum del Chimico, Terza edizione*. Casa Editrice G. Lavagnolo dei Mestieri e delle Professioni dell Industrie e Scienze, Torino, page 37.

Table 11 – Fuming sulfuric acid (oleum) grades

Oleum grade	Mass percent free SO_3	Mass percent total SO_3	Coefficient $H_2SO_4.xSO_3$	H_2SO_4 equivalent (%)	Specific gravity (35°/15°)	Density 20°C (kg/m³)	Freezing point (°C)
10% oleum	10	48.12	0.136	102.3	1.8800	1,878.3	10.4
20% oleum	20	51.60	0.306	104.5	1.8919	1,890.0	11.0
25% oleum	25	53.48	0.408	105.6	1.9350	1,933.2	14.0
30% oleum	30	55.45	0.525	106.8	1.9318	1,930.0	28.0
40% oleum	40	59.73	0.817	109.0	1.9584	1,956.6	32.0
Pyrosulfuric acid	44.94	62.02	1.000	110.1	1.9610	1,935.0	36.01
60% oleum	60	69.85	1.838	113.5	1.9738	1,972.0	
65% oleum	65	72.78	2.275	114.6	1.9938	1,992.0	
Notes: mass density of pure water $\rho(15°C) = 999.090$ kg/m³, and $\rho(20°C) = 998.204$ kg/m³.							

5 Physical Properties of Sulfuric Acid and Oleums

5.1 Mass Density

5.1.1 Density vs. Mass Percentage and Temperature

The mass density of pure aqueous solution of sulfuric acid at a given temperature is an important property that allows the accurate determination of the mass percentage of acid.

Actually, there is a direct relationship existing between the density and the mass percentage of an aqueous solution of sulfuric acid which is valid up to a maximum mass percentage of 96 wt.% H_2SO_4 as exemplified in the following description.

The most accurate and internationally adopted values for the mass density of aqueous solutions of sulfuric acid at atmospheric pressure, denoted by the Greek letter ρ_A, and expressed in kg/m^3, as a function of the mass percentage of sulfuric acid in wt.% H_2SO_4, and for several selected temperatures ranging from 15°C and 80°C along with corresponding degrees Baumé are reported in Table 12.

Table 12 – Mass density and degrees Baumé of sulfuric acid solutions

wt.%	15°C	20°C	25°C	30°C	40°C	50°C	60°C	80°C
1	1,006.0	1,005.1	1,003.8	1,002.2	998.6	994.4	989.5	977.9
	0.86°Bé	0.74°Bé	0.55°Bé	0.32°Bé	-0.20°Bé	-0.82°Bé	-1.54°Bé	-3.28°Bé
2	1,012.9	1,011.8	1,010.4	1,008.7	1,005.0	1,000.6	995.6	983.9
	1.85°Bé	1.69°Bé	1.49°Bé	1.25°Bé	0.72°Bé	0.09°Bé	-0.64°Bé	-2.37°Bé
3	1,019.7	1,018.4	1,016.9	1,015.2	1,011.3	1,006.7	1,001.7	990.0
	2.80°Bé	2.62°Bé	2.41°Bé	2.17°Bé	1.62°Bé	0.97°Bé	0.25°Bé	-1.46°Bé
4	1,026.4	1,025.0	1,023.4	1,021.6	1,017.6	1,012.9	1,007.8	996.1
	3.73°Bé	3.54°Bé	3.32°Bé	3.07°Bé	2.51°Bé	1.85°Bé	1.12°Bé	-0.57°Bé
5	1,033.2	1,031.7	1,030.0	1,028.1	1,024.0	1,019.2	1,014.0	1,002.2

wt.%	15°C	20°C	25°C	30°C	40°C	50°C	60°C	80°C
	4.66°Bé	4.46°Bé	4.22°Bé	3.96°Bé	3.40°Bé	2.73°Bé	2.00°Bé	0.32°Bé
6	1,040.0	1,038.5	1,036.7	1,034.7	1,030.5	1,025.6	1,020.3	1,008.4
	5.58°Bé	5.38°Bé	5.13°Bé	4.86°Bé	4.29°Bé	3.62°Bé	2.88°Bé	1.21°Bé
7	1,046.9	1,045.3	1,043.4	1,041.4	1,037.1	1,032.1	1,026.6	1,014.6
	6.50°Bé	6.28°Bé	6.03°Bé	5.76°Bé	5.19°Bé	4.51°Bé	3.76°Bé	2.09°Bé
8	1,053.9	1,052.2	1,050.2	1,048.1	1,043.7	1,038.6	1,033.0	1,020.9
	7.42°Bé	7.19°Bé	6.93°Bé	6.65°Bé	6.07°Bé	5.39°Bé	4.63°Bé	2.97°Bé
9	1,061.0	1,059.1	1,057.1	1,054.9	1,050.3	1,045.1	1,039.5	1,027.3
	8.34°Bé	8.09°Bé	7.83°Bé	7.55°Bé	6.94°Bé	6.26°Bé	5.51°Bé	3.85°Bé
10	1,068.1	1,066.1	1,064.0	1,061.7	1,057.0	1,051.7	1,046.0	1,033.8
	9.24°Bé	8.99°Bé	8.72°Bé	8.43°Bé	7.82°Bé	7.13°Bé	6.38°Bé	4.74°Bé
11	1,075.3	1,073.1	1,071.0	1,068.6	1,063.7	1,058.4	1,052.6	1,040.3
	10.15°Bé	9.88°Bé	9.61°Bé	9.31°Bé	8.68°Bé	8.00°Bé	7.25°Bé	5.62°Bé
12	1,082.5	1,080.2	1,078.0	1,075.6	1,070.5	1,065.1	1,059.3	1,046.9
	11.05°Bé	10.77°Bé	10.49°Bé	10.19°Bé	9.55°Bé	8.86°Bé	8.12°Bé	6.50°Bé
13	1,089.8	1,087.4	1,085.1	1,082.6	1,077.4	1,071.9	1,066.1	1,053.6
	11.95°Bé	11.65°Bé	11.37°Bé	11.06°Bé	10.42°Bé	9.73°Bé	8.99°Bé	7.38°Bé
14	1,097.1	1,094.7	1,092.2	1,089.7	1,084.4	1,078.8	1,072.9	1,060.3
	12.83°Bé	12.54°Bé	12.24°Bé	11.94°Bé	11.29°Bé	10.59°Bé	9.85°Bé	8.25°Bé
15	1,104.5	1,102.0	1,099.4	1,096.8	1,091.4	1,085.7	1,079.8	1,067.1
	13.72°Bé	13.42°Bé	13.11°Bé	12.80°Bé	12.14°Bé	11.45°Bé	10.72°Bé	9.12°Bé
16	1,112.0	1,109.4	1,106.7	1,104.0	1,098.5	1,092.7	1,086.8	1,074.0
	14.60°Bé	14.30°Bé	13.98°Bé	13.66°Bé	13.00°Bé	12.30°Bé	11.58°Bé	9.99°Bé
17	1,119.5	1,116.8	1,114.1	1,111.3	1,105.7	1,099.8	1,093.8	1,080.9
	15.48°Bé	15.16°Bé	14.85°Bé	14.52°Bé	13.86°Bé	13.16°Bé	12.43°Bé	10.85°Bé
18	1,127.1	1,124.3	1,121.5	1,118.7	1,112.9	1,107.0	1,100.9	1,087.9
	16.35°Bé	16.03°Bé	15.71°Bé	15.39°Bé	14.71°Bé	14.02°Bé	13.29°Bé	11.72°Bé
19	1,134.7	1,131.8	1,129.0	1,126.1	1,120.2	1,114.2	1,108.1	1,095.0
	17.21°Bé	16.89°Bé	16.57°Bé	16.24°Bé	15.56°Bé	14.86°Bé	14.15°Bé	12.58°Bé
20	1,142.4	1,139.4	1,136.5	1,133.5	1,127.5	1,121.5	1,115.3	1,102.1
	18.07°Bé	17.74°Bé	17.42°Bé	17.08°Bé	16.40°Bé	15.71°Bé	14.99°Bé	13.43°Bé
21	1,150.1	1,147.1	1,144.1	1,141.0	1,134.9	1,128.8	1,122.6	1,109.3
	18.92°Bé	18.59°Bé	18.26°Bé	17.92°Bé	17.24°Bé	16.55°Bé	15.84°Bé	14.29°Bé
22	1,157.9	1,154.8	1,151.7	1,148.6	1,142.4	1,136.2	1,129.9	1,116.6

wt.%	15°C	20°C	25°C	30°C	40°C	50°C	60°C	80°C
	19.77°Bé	19.44°Bé	19.10°Bé	18.76°Bé	18.07°Bé	17.38°Bé	16.67°Bé	15.14°Bé
23	1,165.7	1,162.6	1,159.4	1,156.3	1,150.0	1,143.7	1,137.3	1,123.9
	20.61°Bé	20.28°Bé	19.94°Bé	19.60°Bé	18.91°Bé	18.22°Bé	17.51°Bé	15.98°Bé
24	1,173.6	1,170.4	1,167.2	1,164.0	1,157.6	1,151.2	1,144.8	1,131.3
	21.45°Bé	21.11°Bé	20.77°Bé	20.43°Bé	19.74°Bé	19.04°Bé	18.34°Bé	16.83°Bé
25	1,181.6	1,178.3	1,175.0	1,171.8	1,165.3	1,158.8	1,152.3	1,138.8
	22.29°Bé	21.94°Bé	21.60°Bé	21.26°Bé	20.57°Bé	19.87°Bé	19.16°Bé	17.67°Bé
26	1,189.6	1,186.2	1,182.9	1,179.6	1,173.0	1,166.5	1,159.9	1,146.3
	23.11°Bé	22.76°Bé	22.42°Bé	22.08°Bé	21.39°Bé	20.70°Bé	19.99°Bé	18.51°Bé
27	1,197.6	1,194.2	1,190.9	1,187.5	1,180.8	1,174.2	1,167.6	1,153.9
	23.92°Bé	23.58°Bé	23.24°Bé	22.89°Bé	22.20°Bé	21.51°Bé	20.81°Bé	19.34°Bé
28	1,205.7	1,202.3	1,198.9	1,195.5	1,188.7	1,182.0	1,175.3	1,161.6
	24.74°Bé	24.40°Bé	24.06°Bé	23.71°Bé	23.02°Bé	22.33°Bé	21.63°Bé	20.17°Bé
29	1,213.8	1,210.4	1,206.9	1,203.5	1,196.6	1,189.8	1,183.1	1,169.3
	25.54°Bé	25.20°Bé	24.86°Bé	24.52°Bé	23.82°Bé	23.13°Bé	22.44°Bé	20.99°Bé
30	1,222.0	1,218.5	1,215.0	1,211.5	1,204.6	1,197.7	1,190.9	1,177.1
	26.34°Bé	26.00°Bé	25.66°Bé	25.31°Bé	24.63°Bé	23.93°Bé	23.24°Bé	21.82°Bé
31	1,230.2	1,226.7	1,223.2	1,219.6	1,212.6	1,205.7	1,198.8	1,184.9
	27.13°Bé	26.80°Bé	26.46°Bé	26.11°Bé	25.42°Bé	24.74°Bé	24.05°Bé	22.63°Bé
32	1,238.5	1,234.9	1,231.4	1,227.8	1,220.7	1,213.7	1,206.8	1,192.8
	27.92°Bé	27.58°Bé	27.25°Bé	26.90°Bé	26.22°Bé	25.53°Bé	24.85°Bé	23.44°Bé
33	1,246.8	1,243.2	1,239.6	1,236.0	1,228.9	1,221.8	1,214.8	1,200.8
	28.70°Bé	28.37°Bé	28.03°Bé	27.69°Bé	27.01°Bé	26.32°Bé	25.64°Bé	24.25°Bé
34	1,255.2	1,251.5	1,247.9	1,244.3	1,237.1	1,230.0	1,222.9	1,208.8
	29.48°Bé	29.14°Bé	28.80°Bé	28.47°Bé	27.79°Bé	27.11°Bé	26.43°Bé	25.05°Bé
35	1,263.6	1,259.9	1,256.3	1,252.6	1,245.4	1,238.3	1,231.1	1,216.9
	30.25°Bé	29.91°Bé	29.58°Bé	29.24°Bé	28.57°Bé	27.90°Bé	27.22°Bé	25.84°Bé
36	1,272.0	1,268.4	1,264.7	1,261.0	1,253.8	1,246.6	1,239.4	1,225.1
	31.01°Bé	30.68°Bé	30.35°Bé	30.01°Bé	29.35°Bé	28.68°Bé	28.01°Bé	26.64°Bé
37	1,280.5	1,276.9	1,273.2	1,269.5	1,262.2	1,255.0	1,247.7	1,233.4
	31.76°Bé	31.44°Bé	31.11°Bé	30.78°Bé	30.12°Bé	29.46°Bé	28.79°Bé	27.44°Bé
38	1,289.1	1,285.5	1,281.8	1,278.0	1,270.7	1,263.5	1,256.1	1,241.8
	32.52°Bé	32.20°Bé	31.88°Bé	31.54°Bé	30.89°Bé	30.24°Bé	29.56°Bé	28.23°Bé
39	1,297.8	1,294.1	1,290.4	1,286.6	1,279.3	1,272.0	1,264.6	1,250.3

wt.%	15°C	20°C	25°C	30°C	40°C	50°C	60°C	80°C
	33.27°Bé	32.95°Bé	32.63°Bé	32.30°Bé	31.66°Bé	31.01°Bé	30.34°Bé	29.03°Bé
40	1,306.5	1,302.8	1,299.1	1,295.3	1,288.0	1,280.6	1,273.2	1,258.9
	34.02°Bé	33.70°Bé	33.38°Bé	33.06°Bé	32.42°Bé	31.77°Bé	31.11°Bé	29.82°Bé
41	1,315.3	1,311.6	1,307.9	1,304.1	1,296.7	1,289.3	1,281.9	1,267.5
	34.76°Bé	34.45°Bé	34.14°Bé	33.81°Bé	33.18°Bé	32.54°Bé	31.89°Bé	30.60°Bé
42	1,324.2	1,320.5	1,316.7	1,312.9	1,305.5	1,298.1	1,290.7	1,276.2
	35.50°Bé	35.19°Bé	34.88°Bé	34.56°Bé	33.93°Bé	33.30°Bé	32.66°Bé	31.38°Bé
43	1,333.2	1,329.4	1,325.6	1,321.8	1,314.4	1,307.0	1,299.6	1,285.0
	36.24°Bé	35.93°Bé	35.62°Bé	35.30°Bé	34.68°Bé	34.06°Bé	33.43°Bé	32.16°Bé
44	1,342.3	1,338.4	1,334.6	1,330.8	1,323.4	1,316.0	1,308.6	1,293.9
	36.98°Bé	36.66°Bé	36.35°Bé	36.04°Bé	35.43°Bé	34.82°Bé	34.19°Bé	32.94°Bé
45	1,351.5	1,347.6	1,343.7	1,339.9	1,332.5	1,325.1	1,317.7	1,302.9
	37.71°Bé	37.40°Bé	37.09°Bé	36.78°Bé	36.18°Bé	35.57°Bé	34.96°Bé	33.71°Bé
46	1,360.8	1,356.9	1,353.0	1,349.2	1,341.7	1,334.3	1,326.9	1,312.0
	38.45°Bé	38.14°Bé	37.83°Bé	37.53°Bé	36.93°Bé	36.33°Bé	35.72°Bé	34.48°Bé
47	1,370.2	1,366.3	1,362.4	1,358.6	1,351.0	1,343.5	1,336.2	1,321.2
	39.18°Bé	38.87°Bé	38.57°Bé	38.27°Bé	37.67°Bé	37.07°Bé	36.48°Bé	35.25°Bé
48	1,379.7	1,375.8	1,371.9	1,368.0	1,360.4	1,352.8	1,345.5	1,330.5
	39.90°Bé	39.61°Bé	39.31°Bé	39.01°Bé	38.41°Bé	37.81°Bé	37.23°Bé	36.02°Bé
49	1,389.3	1,385.4	1,381.4	1,377.5	1,369.9	1,362.3	1,354.9	1,339.9
	40.63°Bé	40.34°Bé	40.03°Bé	39.74°Bé	39.15°Bé	38.56°Bé	37.98°Bé	36.78°Bé
50	1,399.0	1,395.1	1,391.1	1,387.2	1,379.5	1,371.9	1,364.4	1,349.4
	41.35°Bé	41.06°Bé	40.77°Bé	40.47°Bé	39.89°Bé	39.31°Bé	38.73°Bé	37.54°Bé
51	1,408.8	1,404.9	1,400.9	1,397.0	1,389.3	1,381.6	1,374.0	1,359.0
	42.08°Bé	41.79°Bé	41.50°Bé	41.21°Bé	40.63°Bé	40.05°Bé	39.47°Bé	38.30°Bé
52	1,418.8	1,414.8	1,410.9	1,406.9	1,399.1	1,391.4	1,383.7	1,368.7
	42.80°Bé	42.51°Bé	42.23°Bé	41.94°Bé	41.36°Bé	40.79°Bé	40.21°Bé	39.06°Bé
53	1,428.9	1,424.8	1,420.9	1,416.9	1,409.1	1,401.3	1,393.6	1,378.5
	43.52°Bé	43.23°Bé	42.95°Bé	42.66°Bé	42.10°Bé	41.52°Bé	40.95°Bé	39.81°Bé
54	1,439.1	1,435.0	1,431.0	1,427.0	1,419.1	1,411.3	1,403.6	1,388.4
	44.24°Bé	43.95°Bé	43.67°Bé	43.39°Bé	42.82°Bé	42.26°Bé	41.69°Bé	40.56°Bé
55	1,449.4	1,445.3	1,441.2	1,437.2	1,429.3	1,421.4	1,413.7	1,398.4
	44.96°Bé	44.67°Bé	44.39°Bé	44.11°Bé	43.55°Bé	42.99°Bé	42.43°Bé	41.31°Bé
56	1,459.8	1,455.7	1,451.6	1,447.5	1,439.6	1,431.7	1,423.9	1,408.5

wt.%	15°C	20°C	25°C	30°C	40°C	50°C	60°C	80°C
	45.67°Bé	45.39°Bé	45.11°Bé	44.83°Bé	44.28°Bé	43.72°Bé	43.17°Bé	42.05°Bé
57	1,470.3	1,466.2	1,462.1	1,458.0	1,450.0	1,442.0	1,434.2	1,418.7
	46.38°Bé	46.10°Bé	45.83°Bé	45.55°Bé	45.00°Bé	44.45°Bé	43.90°Bé	42.79°Bé
58	1,480.9	1,476.8	1,472.6	1,468.5	1,460.4	1,452.4	1,444.6	1,429.0
	47.09°Bé	46.81°Bé	46.53°Bé	46.26°Bé	45.71°Bé	45.17°Bé	44.63°Bé	43.53°Bé
59	1,491.6	1,487.5	1,483.2	1,479.1	1,470.9	1,462.9	1,455.1	1,439.3
	47.79°Bé	47.52°Bé	47.24°Bé	46.97°Bé	46.42°Bé	45.88°Bé	45.35°Bé	44.26°Bé
60	1,502.4	1,498.3	1,494.0	1,489.8	1,481.6	1,473.5	1,465.6	1,449.7
	48.49°Bé	48.22°Bé	47.95°Bé	47.67°Bé	47.13°Bé	46.59°Bé	46.06°Bé	44.98°Bé
61	1,513.3	1,509.1	1,504.8	1,500.6	1,492.3	1,484.2	1,476.2	1,460.2
	49.18°Bé	48.92°Bé	48.64°Bé	48.37°Bé	47.83°Bé	47.30°Bé	46.77°Bé	45.70°Bé
62	1,524.3	1,520.0	1,515.7	1,511.5	1,503.1	1,495.0	1,486.9	1,470.8
	49.87°Bé	49.61°Bé	49.33°Bé	49.07°Bé	48.53°Bé	48.01°Bé	47.48°Bé	46.41°Bé
63	1,535.4	1,531.0	1,526.7	1,522.5	1,514.0	1,505.8	1,497.7	1,481.5
	50.56°Bé	50.29°Bé	50.02°Bé	49.76°Bé	49.23°Bé	48.71°Bé	48.18°Bé	47.13°Bé
64	1,546.5	1,542.1	1,537.8	1,533.5	1,525.0	1,516.7	1,508.6	1,492.3
	51.24°Bé	50.97°Bé	50.71°Bé	50.45°Bé	49.92°Bé	49.40°Bé	48.88°Bé	47.83°Bé
65	1,557.8	1,553.3	1,549.0	1,544.6	1,536.1	1,527.7	1,519.5	1,503.1
	51.92°Bé	51.65°Bé	51.39°Bé	51.12°Bé	50.61°Bé	50.09°Bé	49.57°Bé	48.53°Bé
66	1,569.1	1,564.6	1,560.2	1,555.8	1,547.2	1,538.8	1,530.5	1,514.0
	52.59°Bé	52.32°Bé	52.06°Bé	51.80°Bé	51.28°Bé	50.77°Bé	50.26°Bé	49.23°Bé
67	1,580.5	1,576.0	1,571.5	1,567.1	1,558.4	1,549.9	1,541.6	1,524.9
	53.26°Bé	52.99°Bé	52.73°Bé	52.47°Bé	51.96°Bé	51.45°Bé	50.94°Bé	49.91°Bé
68	1,592.0	1,587.4	1,582.9	1,578.5	1,569.7	1,561.1	1,552.8	1,535.9
	53.92°Bé	53.66°Bé	53.40°Bé	53.14°Bé	52.63°Bé	52.12°Bé	51.62°Bé	50.59°Bé
69	1,603.5	1,598.9	1,594.4	1,589.9	1,581.1	1,572.4	1,564.0	1,547.0
	54.57°Bé	54.31°Bé	54.06°Bé	53.80°Bé	53.29°Bé	52.78°Bé	52.29°Bé	51.27°Bé
70	1,615.1	1,610.5	1,605.9	1,601.4	1,592.5	1,583.8	1,575.3	1,558.2
	55.22°Bé	54.97°Bé	54.71°Bé	54.45°Bé	53.95°Bé	53.45°Bé	52.95°Bé	51.94°Bé
71	1,626.8	1,622.1	1,617.5	1,613.0	1,604.0	1,595.2	1,586.7	1,569.4
	55.87°Bé	55.61°Bé	55.36°Bé	55.11°Bé	54.60°Bé	54.10°Bé	53.62°Bé	52.61°Bé
72	1,638.5	1,633.8	1,629.2	1,624.6	1,615.5	1,606.7	1,598.1	1,580.6
	56.50°Bé	56.25°Bé	56.00°Bé	55.75°Bé	55.24°Bé	54.75°Bé	54.27°Bé	53.26°Bé
73	1,650.3	1,645.6	1,640.9	1,636.3	1,627.1	1,618.2	1,609.5	1,591.9

wt.%	15°C	20°C	25°C	30°C	40°C	50°C	60°C	80°C
	57.14°Bé	56.89°Bé	56.63°Bé	56.39°Bé	55.88°Bé	55.39°Bé	54.91°Bé	53.91°Bé
74	1,662.2	1,657.4	1,652.6	1,648.0	1,638.7	1,629.7	1,620.9	1,603.1
	57.77°Bé	57.51°Bé	57.26°Bé	57.01°Bé	56.52°Bé	56.03°Bé	55.54°Bé	54.55°Bé
75	1,674.0	1,669.2	1,664.4	1,659.7	1,650.3	1,641.2	1,632.2	1,614.2
	58.38°Bé	58.13°Bé	57.88°Bé	57.63°Bé	57.14°Bé	56.65°Bé	56.16°Bé	55.17°Bé
76	1,685.8	1,681.0	1,676.1	1,671.3	1,661.9	1,652.6	1,643.5	1,625.2
	58.99°Bé	58.74°Bé	58.49°Bé	58.24°Bé	57.75°Bé	57.26°Bé	56.77°Bé	55.78°Bé
77	1,697.6	1,692.7	1,687.8	1,682.9	1,673.4	1,664.0	1,654.7	1,636.1
	59.59°Bé	59.34°Bé	59.09°Bé	58.84°Bé	58.35°Bé	57.86°Bé	57.37°Bé	56.37°Bé
78	1,709.3	1,704.3	1,699.4	1,694.4	1,684.7	1,675.1	1,665.7	1,646.9
	60.17°Bé	59.92°Bé	59.68°Bé	59.42°Bé	58.93°Bé	58.44°Bé	57.95°Bé	56.96°Bé
79	1,720.9	1,715.8	1,710.8	1,705.8	1,695.9	1,686.2	1,676.6	1,657.5
	60.74°Bé	60.49°Bé	60.24°Bé	60.00°Bé	59.50°Bé	59.01°Bé	58.52°Bé	57.52°Bé
80	1,732.3	1,727.2	1,722.1	1,717.0	1,706.9	1,697.1	1,687.3	1,668.0
	61.30°Bé	61.05°Bé	60.80°Bé	60.55°Bé	60.05°Bé	59.56°Bé	59.06°Bé	58.07°Bé
81	1,743.5	1,738.3	1,733.1	1,727.9	1,717.7	1,707.7	1,697.8	1,678.2
	61.83°Bé	61.59°Bé	61.33°Bé	61.08°Bé	60.58°Bé	60.09°Bé	59.60°Bé	58.60°Bé
82	1,754.4	1,749.1	1,743.7	1,738.5	1,728.1	1,718.0	1,708.0	1,688.2
	62.35°Bé	62.10°Bé	61.84°Bé	61.59°Bé	61.09°Bé	60.60°Bé	60.11°Bé	59.11°Bé
83	1,764.9	1,759.4	1,754.0	1,748.7	1,738.2	1,727.9	1,717.9	1,697.9
	62.84°Bé	62.59°Bé	62.33°Bé	62.08°Bé	61.58°Bé	61.08°Bé	60.59°Bé	59.60°Bé
84	1,774.8	1,769.3	1,763.9	1,758.5	1,747.9	1,737.5	1,727.4	1,707.2
	63.30°Bé	63.05°Bé	62.80°Bé	62.54°Bé	62.04°Bé	61.55°Bé	61.06°Bé	60.07°Bé
85	1,784.1	1,778.6	1,773.2	1,767.8	1,757.1	1,746.6	1,736.4	1,716.1
	63.73°Bé	63.48°Bé	63.23°Bé	62.98°Bé	62.48°Bé	61.98°Bé	61.49°Bé	60.51°Bé
86	1,792.7	1,787.2	1,781.8	1,776.3	1,765.7	1,755.2	1,744.9	1,724.5
	64.12°Bé	63.87°Bé	63.62°Bé	63.37°Bé	62.88°Bé	62.39°Bé	61.90°Bé	60.92°Bé
87	1,800.6	1,795.1	1,789.7	1,784.2	1,773.6	1,763.2	1,752.9	1,732.4
	64.47°Bé	64.22°Bé	63.98°Bé	63.73°Bé	63.25°Bé	62.76°Bé	62.28°Bé	61.30°Bé
88	1,807.7	1,802.2	1,796.8	1,791.4	1,780.9	1,770.5	1,760.2	1,739.7
	64.79°Bé	64.54°Bé	64.30°Bé	64.06°Bé	63.58°Bé	63.10°Bé	62.62°Bé	61.65°Bé
89	1,814.1	1,808.7	1,803.3	1,797.9	1,787.4	1,777.0	1,766.9	1,746.4
	65.07°Bé	64.83°Bé	64.59°Bé	64.35°Bé	63.88°Bé	63.40°Bé	62.94°Bé	61.97°Bé
90	1,819.8	1,814.4	1,809.1	1,803.8	1,793.3	1,782.9	1,772.9	1,752.5

wt.%	15°C	20°C	25°C	30°C	40°C	50°C	60°C	80°C
	65.32°Bé	65.08°Bé	64.85°Bé	64.61°Bé	64.14°Bé	63.67°Bé	63.21°Bé	62.26°Bé
91	1,824.8	1,819.5	1,814.2	1,809.0	1,798.6	1,788.3	1,778.3	1,758.1
	65.54°Bé	65.31°Bé	65.07°Bé	64.85°Bé	64.38°Bé	63.92°Bé	63.46°Bé	62.52°Bé
92	1,829.3	1,824.0	1,818.8	1,813.6	1,803.3	1,793.2	1,783.2	1,763.3
	65.73°Bé	65.50°Bé	65.28°Bé	65.05°Bé	64.59°Bé	64.14°Bé	63.69°Bé	62.77°Bé
93	1,833.1	1,827.9	1,822.7	1,817.6	1,807.4	1,797.4	1,787.6	1,768.1
	65.90°Bé	65.67°Bé	65.45°Bé	65.22°Bé	64.77°Bé	64.33°Bé	63.89°Bé	62.99°Bé
94	1,836.3	1,831.2	1,826.0	1,821.0	1,810.9	1,801.1	1,791.4	
	66.04°Bé	65.82°Bé	65.59°Bé	65.37°Bé	64.93°Bé	64.49°Bé	64.06°Bé	
95	1,838.8	1,833.7	1,828.6	1,823.6	1,813.7	1,804.0	1,794.4	
	66.14°Bé	65.92°Bé	65.70°Bé	65.49°Bé	65.05°Bé	64.62°Bé	64.19°Bé	
96	1,840.6	1,835.5	1,830.5	1,825.5	1,815.7	1,806.0	1,796.5	
	66.22°Bé	66.00°Bé	65.79°Bé	65.57°Bé	65.14°Bé	64.71°Bé	64.29°Bé	
97	1,841.4	1,836.4	1,831.4	1,826.4	1,816.6	1,807.1	1,797.7	
	66.26°Bé	66.04°Bé	65.83°Bé	65.61°Bé	65.18°Bé	64.76°Bé	64.34°Bé	
98	1,841.1	1,836.1	1,831.0	1,826.1	1,816.3	1,806.8	1,797.6	
	66.24°Bé	66.03°Bé	65.81°Bé	65.60°Bé	65.17°Bé	64.75°Bé	64.34°Bé	
99	1,839.3	1,834.2	1,829.2	1,824.2	1,814.5	1,805.0	1,795.8	
	66.17°Bé	65.95°Bé	65.73°Bé	65.51°Bé	65.09°Bé	64.67°Bé	64.26°Bé	
100	1,835.7	1,830.5	1,825.5	1,820.5	1,810.7	1,801.3	1,792.2	
	66.01°Bé	65.79°Bé	65.57°Bé	65.35°Bé	64.92°Bé	64.50°Bé	64.09°Bé	

5.1.2   Determination of Concentration by Densitometry

From the previous tabulated data, when the actual mass density ρ_A of an aqueous solution of sulfuric acid of unknown strength, w_A, is measured at a given temperature t_A, the determination of the mass percentage can be calculated with great accuracy by performing a double linear interpolation described hereafter.

First, we need to locate in Table 12 the closest lower (t_1) and higher (t_2) temperatures surrounding the previous measured temperature, then the lower and higher mass densities than ρ_A, for these two temperatures ρ_1 and ρ_2, and their respective mass percentages w_1, and w_2. Thus after arranging them in the following matrix it is possible to calculate the unknown strength:

Mass percentages	t_1	t_A	t_2
w_1	$\rho_1(t_1)$	$\rho_1(t_A)$	$\rho_1(t_2)$
w_A	–	$\rho_A(t_A)$	–
w_2	$\rho_2(t_1)$	$\rho_2(t_A)$	$\rho_2(t_2)$

The lower and higher mass densities at the temperature t_A are then obtained simply by linear interpolation from the two sets of data listed below:

t_1	$\rho_1(t_1)$	t_1	$\rho_1(t_1)$
t_A	$\rho_1(t_A)$	t_A	$\rho_1(t_A)$
t_2	$\rho_1(t_2)$	t_2	$\rho_1(t_2)$

This yields the two densities:

$$\rho_1(t_A) = \rho_1(t_1) + [\rho_1(t_2) - \rho_1(t_1)][(t_A - t_1)/ (t_2 - t_1)]$$

$$\rho_2(t_A) = \rho_2(t_1) + [\rho_2(t_2) - \rho_2(t_1)][(t_A - t_1)/ (t_2 - t_1)]$$

Then from the two above interpolated mass densities, the mass percentage is obtained again by performing a second linear interpolation using the following set:

w_1	$\rho_1(t_A)$
w_A	$\rho_A(t_A)$
w_2	$\rho_2(t_A)$

This yields the mass percentage as follows:

$$w_A = w_1 + (w_2 - w_1)[\rho_A(t_A) - \rho_1(t_A)]/ [\rho_2(t_A) - \rho_1(t_A)]$$

Example – If the mass density of an aqueous solution of sulfuric acid measured with a Mohr-Westphal hydrostatic balance is 1,308.4 kg/m³ when measured at 22.3°C. From the tabulated data the closest values are:

Mass percentages	20°C	22.3°C	25°C
w_1 = 41 wt.%	1,302.8	$\rho_1(22.3)$	1,299.1
w_A	-	**1,308.4**	-
w_2 = 42 wt.%	1,311.6	$\rho_2(22.3)$	1,307.9

Thus from the first linear interpolation, the lower and upper mass densities at 22.3°C are:

$$\rho_1(22.3°C) = 1{,}301.1 \ kg/m^3$$

$$\rho_2(22.3°C) = 1{,}309.9 \ kg/m^3$$

Then, the mass percentage is obtained again by performing a second linear interpolation using the following table:

41 wt.%	1,301.1
w_A	1,308.4
42 wt.%	1,309.9

Thus the unknown acid strength is 41.83 wt.% H_2SO_4.

Later in this section we will see that combining the measurement of the mass density along with a determination of a second physical quantity such as the sound velocity or the refractive index will allow

5.2   Dynamic Viscosity

The dynamic or absolute viscosity in mPa.s (cP) of aqueous solutions of sulfuric acid vs. the mass percentage of acid and the temperature is depicted in Figure 6 from the experimental data of Rhodes and Barbour [45]

---

[45] RHODES; F.H.; and BARBOUR, C.B. (1923) The viscosities of mixtures of sulfuric acid and water. *Industrial and Engineering Chemistry* **15**(8)850-852.

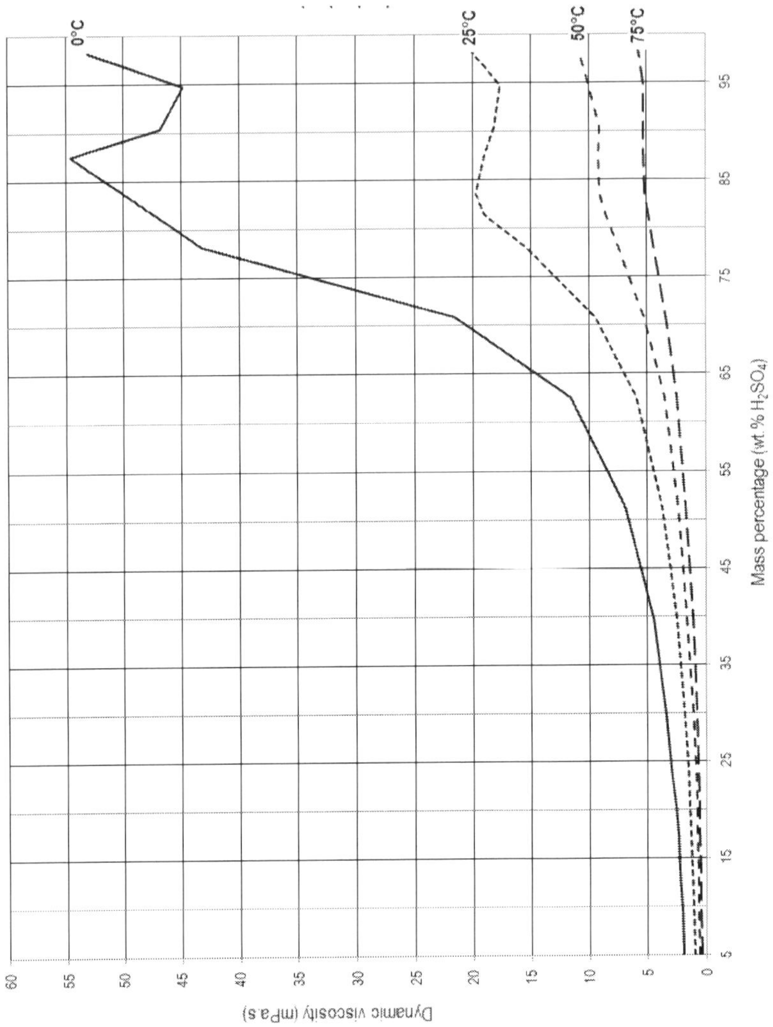

Figure 6 – Dynamic viscosity of sulfuric acid vs. mass percent and temperature

5.3 Sound Velocity

The velocity of sound in sulfuric acid and oleum can be used to determine along with the measure of density the sulfuric acid concentration.

Table 13 – Sound velocity and mass density vs. mass percentage

Mass percentage (wt.%)	Mass density (kg/m³)	Sound velocity (m/s)	Mass percentage (wt.%)	Mass density (kg/m³)	Sound velocity (m/s)
0	998.2	1,482	60	1,498.3	1,563
5	1,031.7	1,479	65	1,553.3	1,555
10	1,066.1	1,482	70	1,610.5	1,551
15	1,102.0	1,494	75	1,669.2	1,544
20	1,139.4	1,509	80	1,727.2	1,540
25	1,178.3	1,524	85	1,778.6	1,513
30	1,218.5	1,540	90	1,814.4	1,471
35	1,259.9	1,551	93	1,827.9	1,418
40	1,302.8	1,563	95	1,833.7	1,383
45	1,347.6	1,566	98	1,836.1	1,303
50	1,395.1	1,563	100	1,830.5	1,257
55	1,445.3	1,563			

The variation of sound velocity as a function of the mass percentage of sulfuric acid from 0 to 100 wt.% H_2SO_4 is depicted in Figure 7.

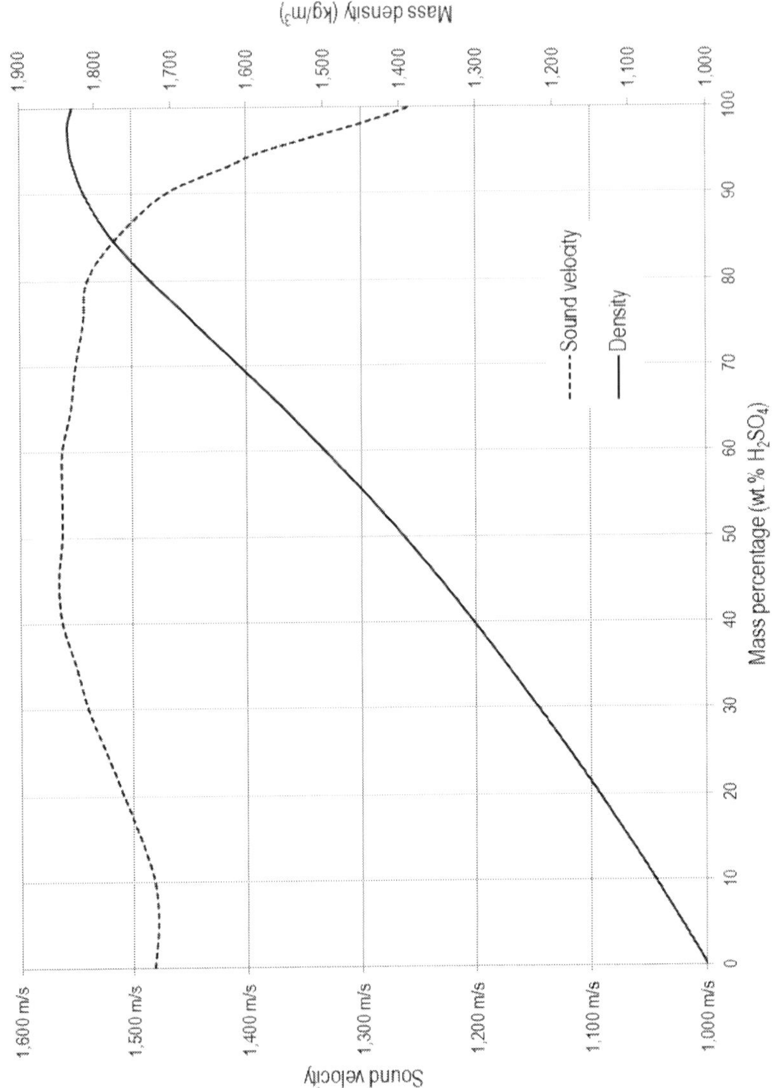

Figure 7 – Sound velocity and mass density vs. mass percent of sulfuric acid at 20°C

Thus the determination of the titer of sulfuric acid by these accurate measurements of the density alone or together with the sound velocity [46], neither

[46] VRAY, D.; BERCHOUZ, D.; DELACHARTRE, P.; and GIMENEZ, G. (1992) Speed of sound in sulfuric acid solution: Application to density measurement. *IEEE 1992 Ultrasonics Symposium Proceedings*, 1992, pp. 969-972 vol.2.

sample preparation (e.g., dilution) nor chemical reagent are required which confers a great simplicity to these measurements.

From the above plot, we can see clearly that for mass percentages ranging from 0 wt.% to 90 wt.% H_2SO_4, the determination of the mass density alone is satisfactory but for mass percentages above 90 wt.% another physical quantity such as sound velocity must be determined to solve the problem.

5.4   Specific Heat Capacity

The variation of the specific heat capacity of aqueous solutions of sulfuric acid vs. the mass percentage of H_2SO_4 (see Figure 8) decreases with increasing concentration of sulfuric acid with a bump occurring between 80°C and 85°C.

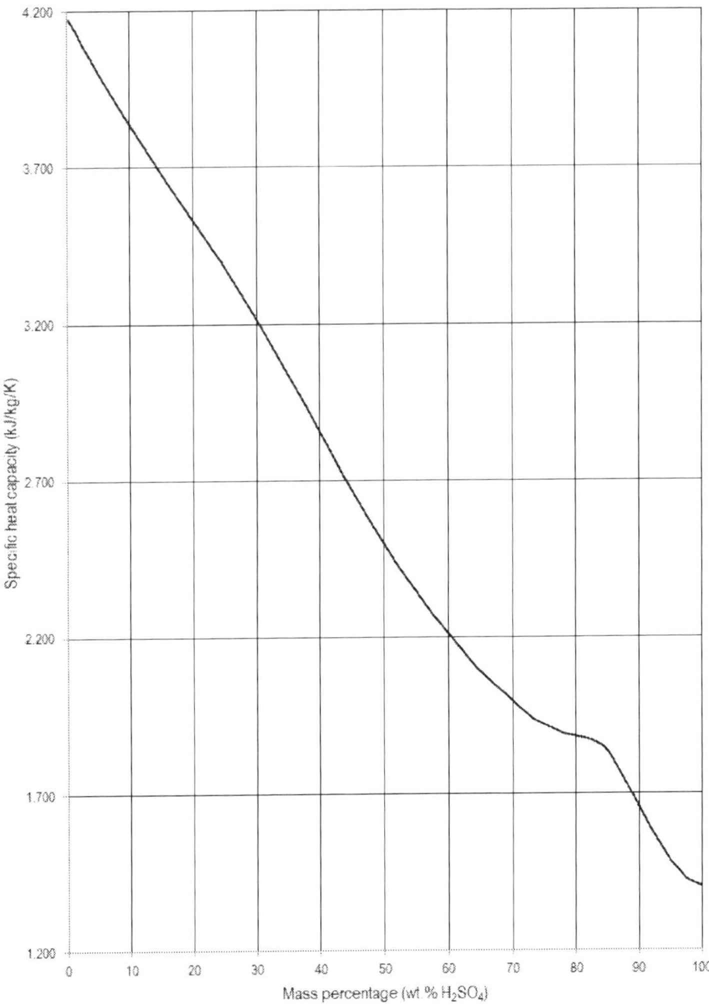

Figure 8 – Specific heat capacity of aqueous solutions of sulfuric acid

5.5 Vapour Pressure

The total vapor pressure of aqueous solutions of sulfuric acid expressed in kPa from data of Abel [47] and later then from Gmitro and Vermeulen [48, 49] are reported in Table 14.

---

[47] ABEL, E. (1948) On the experimental bases for the calculation of the sulfuric acid vapor pressure above the sulfuric acid-water system. *The Journal of Physical Chemistry*, **52**(5) 908-914.

Table 14 – Total vapor pressure vs. mass percentage of sulfuric acid

b.p.	10 w/o 101 °C	20 w/o 103 °C	30 w/o 107 °C	40 w/o 113 °C	50 w/o 123 °C	60 w/o 140 °C	70 w/o 166 °C	75 w/o 184 °C	80 w/o 203 °C	85 w/o 232 °C	90 w/o 265 °C	92 w/o 276 °C	94 w/o 292 °C	96 w/o 310 °C	97 w/o 320 °C	98 w/o 327.5 °C	98.5 w/o 326.8 °C	99 w/o 308 °C	99.5 w/o 290 °C
0 °C	0.58	0.53	0.45	0.33	0.19	3.36E-02	2.07E-02	7.47E-03	1.97E-03	3.43E-04	5.18E-05	2.43E-05	1.09E-05	4.16E-06	2.35E-06	1.17E-06	7.70E-07	4.84E-07	3.31E-07
10 °C	1.17	1.07	0.91	0.67	0.41	0.18	4.67E-02	1.75E-02	4.90E-03	9.52E-04	1.59E-04	7.65E-05	3.48E-05	1.36E-05	7.74E-06	3.92E-06	2.62E-06	1.68E-06	1.21E-06
20 °C	2.23	2.05	1.74	1.30	0.80	0.37	9.95E-02	3.88E-02	1.15E-02	2.45E-03	4.49E-04	2.21E-04	1.02E-04	4.07E-05	2.35E-05	1.21E-05	8.16E-06	5.38E-06	4.08E-06
30 °C	4.04	3.73	3.19	2.41	1.51	0.71	0.20	8.11E-02	2.53E-02	5.89E-03	1.17E-03	5.90E-04	2.79E-04	1.13E-04	6.59E-05	3.45E-05	2.36E-05	1.59E-05	1.27E-05
40 °C	7.03	6.49	5.58	4.27	2.72	1.31	0.39	0.16	5.31E-02	1.34E-02	3.85E-03	1.47E-03	7.08E-04	2.93E-04	1.73E-04	9.17E-05	6.36E-05	4.37E-05	3.67E-05
50 °C	11.70	10.90	9.39	7.25	4.70	2.32	0.72	0.31	0.11	2.86E-02	6.53E-03	3.44E-03	1.69E-03	7.12E-04	4.25E-04	2.29E-04	1.61E-04	1.13E-04	1.00E-04
60 °C	18.90	17.50	15.20	11.90	7.82	3.95	1.27	0.57	0.20	5.84E-02	1.41E-02	7.59E-03	3.80E-03	1.64E-03	9.88E-04	5.41E-04	3.84E-04	2.76E-04	2.57E-04
70 °C	29.60	27.50	23.90	18.80	12.60	6.51	2.17	0.97	0.38	0.11	2.91E-02	1.59E-02	8.13E-03	3.57E-03	2.18E-03	1.21E-03	8.71E-04	6.37E-04	6.25E-04
80 °C	44.90	41.70	36.50	29.00	19.60	10.40	3.60	1.70	0.67	0.21	5.71E-02	3.19E-02	1.66E-02	7.42E-03	4.59E-03	2.59E-03	1.88E-03	1.40E-03	1.46E-03
90 °C	66.40	61.70	54.20	43.40	29.80	16.10	5.78	2.81	1.15	0.38	0.11	6.12E-02	3.24E-02	1.48E-02	9.23E-03	5.29E-03	3.99E-03	2.97E-03	3.23E-03
100 °C	95.70	89.10	78.60	63.40	44.10	24.40	9.05	4.52	1.92	0.67	0.20	0.11	6.07E-02	2.83E-02	1.79E-02	1.04E-02	7.73E-03	6.02E-03	6.89E-03
110 °C	134.90	125.80	111.30	90.40	63.80	36.00	13.80	7.08	3.12	1.12	0.34	0.20	0.11	5.22E-02	3.33E-02	1.97E-02	1.48E-02	1.18E-02	1.42E-02
120 °C	186.30	174.00	154.40	128.40	90.30	51.90	20.60	10.80	4.93	1.83	0.58	0.35	0.19	9.31E-02	6.01E-02	3.60E-02	2.74E-02	2.23E-02	2.82E-02
130 °C	252.40	236.10	210.10	173.20	125.30	73.40	30.10	16.20	7.60	2.91	0.94	0.58	0.33	0.16	0.10	6.39E-02	4.92E-02	4.09E-02	5.43E-02
140 °C	336.10	314.90	281.00	233.30	170.80	102.00	43.10	23.60	11.50	4.51	1.51	0.94	0.54	0.27	0.18	0.11	8.59E-02	7.30E-02	0.10
150 °C	440.40	413.20	369.70	309.00	228.90	139.20	60.50	33.90	17.00	6.83	2.35	1.49	0.87	0.44	0.30	0.18	0.15	0.13	0.18
160 °C	568.50	534.20	479.30	403.10	302.10	187.00	83.70	47.80	24.60	10.10	3.57	2.30	1.36	0.71	0.48	0.30	0.24	0.22	0.33
170 °C	723.60	681.00	612.70	518.50	393.00	247.50	113.80	66.20	35.00	14.70	5.32	3.47	2.08	1.10	0.75	0.48	0.39	0.36	0.57
180 °C	909.30	857.10	773.10	658.40	504.50	323.30	152.50	90.20	48.90	20.90	7.75	5.14	3.12	1.68	1.15	0.76	0.62	0.58	0.96
190 °C	1128.90	1065.80	964.00	825.90	639.70	416.90	201.70	121.20	67.30	29.20	11.10	7.47	4.61	2.51	1.75	1.16	0.97	0.93	1.60
200 °C	1386.10	1310.70	1188.70	1024.50	802.00	531.20	263.30	160.60	91.30	40.20	15.60	10.71	6.66	3.70	2.59	1.75	1.47	1.45	2.59
210 °C	1684.10	1595.10	1450.50	1257.60	994.80	669.60	339.60	210.10	122.10	54.40	21.61	15.01	9.46	5.34	3.78	2.60	2.22	2.24	4.14
220 °C	2026.40	1922.50	1752.90	1528.70	1221.70	835.40	433.10	271.50	161.00	72.60	29.51	20.72	13.23	7.59	5.43	3.78	3.27	3.38	6.50
230 °C	2416.00	2296.00	2099.20	1841.40	1486.40	1032.20	546.60	346.80	209.80	95.60	39.62	28.23	18.25	10.61	7.68	5.42	4.78	5.05	10.01
240 °C	2856.10	2718.80	2492.70	2199.20	1792.90	1264.10	583.20	438.20	270.10	124.21	52.53	37.94	24.78	14.68	10.69	7.68	6.86	7.44	15.22
250 °C	3349.40	3193.90	2936.40	2605.60	2145.20	1535.10	845.90	548.10	343.90	159.41	68.84	50.37	33.23	19.98	14.75	10.74	9.73	10.81	22.75
260 °C	3898.40	3724.00	3433.40	3064.20	2547.20	1849.60	1038.40	679.10	433.20	202.32	88.17	66.11	44.10	26.82	19.99	14.84	13.65	15.57	33.60
270 °C	4505.50	4311.60	3986.50	3578.40	3003.00	2212.20	1264.20	833.70	540.21	254.03	114.21	85.77	57.81	35.74	26.83	17.30	18.89	22.06	49.10
280 °C	5172.60	4959.00	4598.40	4151.40	3516.80	2527.50	1527.20	1014.70	667.31	315.74	144.87	110.17	74.87	46.95	35.62	27.43	25.95	30.93	70.70
290 °C	5901.50	5668.10	5271.50	4786.60	4092.60	3100.40	1831.50	1225.01	817.02	388.66	181.96	140.20	98.09	61.09	46.81	36.63	35.20	43.10	100.60
300 °C	6693.40	6440.70	6008.10	5486.90	4734.70	3636.10	2181.40	1467.51	991.63	474.10	226.50	176.70	122.12	78.71	60.98	48.54	47.43	59.20	141.70
310 °C	7549.50	7278.10	6810.10	6255.30	5447.00	4239.80	2581.21	1745.32	1193.94	573.36	279.69	220.77	153.88	100.57	78.71	63.82	63.20	80.80	197.40
320 °C	8470.50	8181.60	7679.20	7094.70	6233.80	4716.80	3035.51	2061.12	1426.46	687.84	342.57	273.56	192.21	127.45	100.72	83.20	83.50	109.20	272.20
330 °C	9456.70	9151.80	8617.20	8007.80	7099.00	5672.70	3548.92	2418.24	1691.70	818.87	416.56	336.51	238.29	160.15	127.87	107.50	109.60	146.40	372.10
340 °C	10508.30	10189.40	9625.20	8997.00	8046.60	6513.01	4126.22	2819.36	1992.15	967.75	502.91	410.95	293.08	199.70	161.10	138.00	142.60	194.70	504.10
350 °C	11625.10	11294.70	10704.30	10064.70	9080.60	7443.71	4772.34	3267.50	2330.52	1135.92	603.18	498.47	358.08	247.30	201.60	175.90	184.40	257.00	677.30

[48] G MITRO, J.I.; and V ERMEULEN, Th. (1963) Vapor-Liquid Equilibria for Aqueous Sulfuric Acid. Report UCRL-10886, Lawrence Berkeley National Lab. (LBNL), Berkeley, CA, U.S. Department of Energy.

[49] G MITRO, J.I.; and V ERMEULEN, Th. (1964) Vapor-liquid equilibria for aqueous sulfuric acid. *AIChE Journal* **10**(5) pp. 740-746.

5.6 Boiling Point

The boiling point of aqueous solution of sulfuric acid under normal atmospheric pressure (101.325 kPa) increases with the sulfuric acid mass percent up to a peak temperature of 337°C (638.6°F) for 98.5 wt.% H_2SO_4 (Figure 9) which corresponds to the azeotrope composition discussed previously afterwards the boiling point decreases abruptly again down to 290°C for 100 wt.%

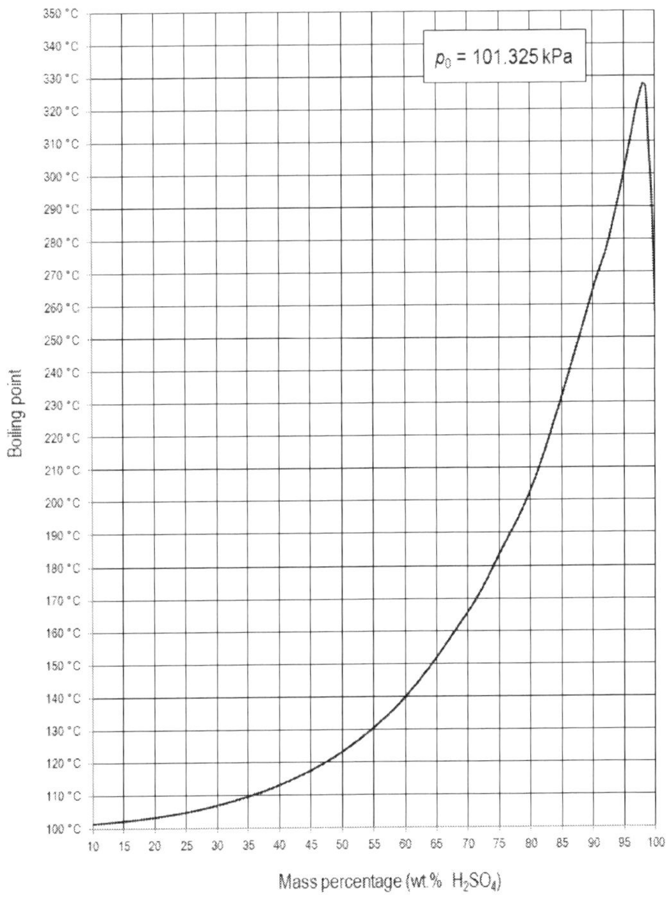

Figure 9 – Boiling point of sulfuric acid solutions

5.7 Vapor Pressure and Dühring Plot

The vapor pressure of aqueous solutions of sulfuric acid is less than that of pure water at the same temperature. Conversely, for a given pressure the boiling point of the solution is higher than that of pure water.

Table 15 – Selected boiling points of sulfuric acid solutions

wt.%	b.p./°C	wt.%	b.p./°C
10	101	90	265
20	103	92	276
30	107	94	292
40	113	96	311
50	123	97	323
60	140	98	334
70	166	**98.5**	**337**
75	184	99	332
80	203	99.5	308
85	232	100	290

The difference between the boiling points of solution vs. that of pure water is referred as the ***boiling point elevation*** (BPE) of the solution.

The empirical Ramsay-Young equation devised experimentally by W. Ramsay and S. Young in 1885 stating that the ratio of the boiling point of two similar chemical substances remains constant at all the vapor pressures no matter the external pressure

$$T_1/T_2 = T_3/T_4$$

By rearranging the above equation:

$$(T_3 - T_1)/(T_4 - T_2) = \text{constant}$$

The latter equation was developed by Eugen Dühring in 1878 for aqueous solutions of inorganic salts named afterward the ***Dühring's rule*** which states that a linear relationship exists between the temperatures at which two aqueous solutions exert the same vapour pressure.

The rule is often used to compare a pure liquid and a solution at a given concentration. Dühring's plot is a graphical representation of such a relationship, typically with the boiling point of water along the abscissa x-axis and the boiling point of the solution along the ordinate y-axis; each line of the graph represents a constant concentration.

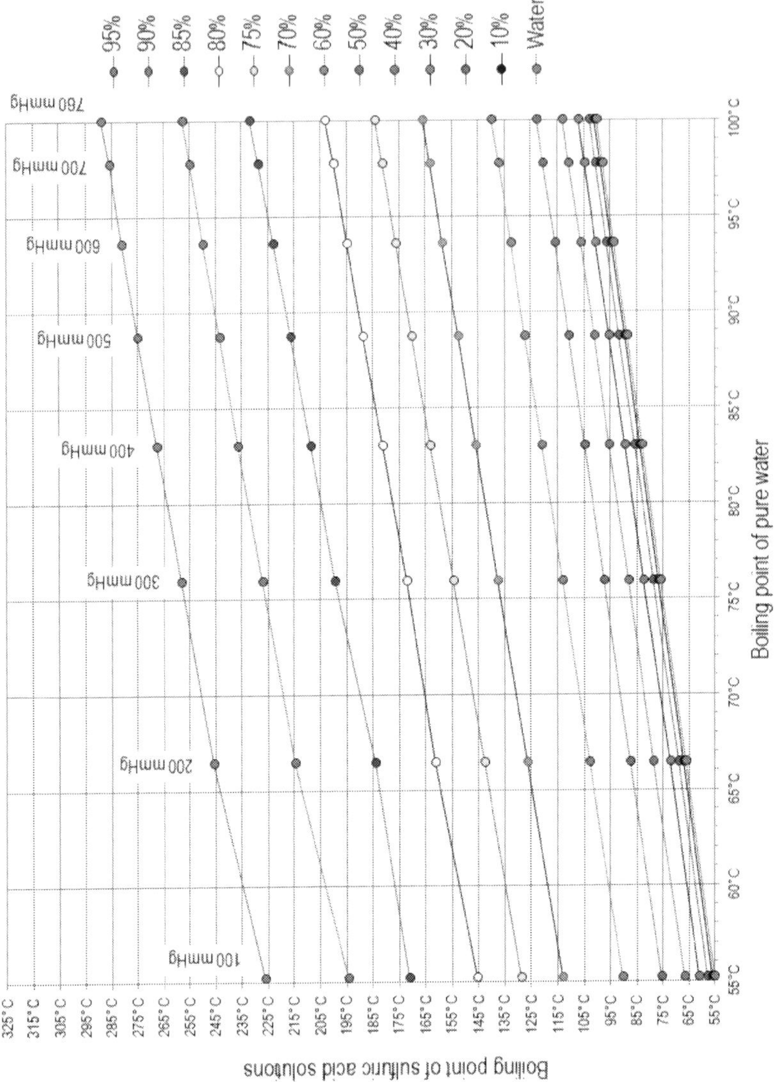

Figure 10 – Dühring plot for aqueous solutions of sulfuric acid

Example – For instance, if we need to determine the boiling point elevation (BPE) of an aqueous solution of sulfuric acid with a mass percentage of 80 wt.% H_2SO_4 at an absolute pressure of 200 mmHg (ca. 26.7 kPa), we need first to retrieve the boiling point of pure water from the international steam tables: 66.4°C. Then drawing a vertical line at that abscissa on the Dühring plot, the intersection with the isoconcentration curve at 70 wt.% indicates that the boiling point of the solution under that reduced pressure is 160°C, thus the $BPE = 60°C$.

5.8   Electrical Conductivity

Because of the strong acidity of the first hydrogen and high dissociation constant, the ***electrical conductivity*** of aqueous solutions of sulfuric acid, denoted κ, is extremely high and sulfuric acid solutions among the most electrically conductive aqueous electrolytes. The electrical conductivity of sulfuric acid solution expressed in S/m as reported in Table 16 hereafter.

Table 16 – Electrical conductivity of sulfuric acid

wt.%	15°C	20°C	25°C	30°C	40°C	50°C	60°C	70°C	80°C	90°C
5	19.66	21.02	22.32	23.51	24.65	26.76	28.53	30.07	31.18	32.50
10	37.70	40.55	43.35	44.55	49.48	53.90	56.85	60.00	62.37	65.13
15	52.00	56.25	60.45	62.20	69.40	75.78	80.15	85.45	88.93	93.13
20	62.65	68.05	73.15	75.70	85.30	91.90	99.80	107.10	111.80	117.68
25	68.80	74.80	80.80	83.73	95.23	105.93	113.30	122.15	128.43	135.78
30	**70.75**	**77.20**	**83.85**	**87.05**	99.73	111.93	120.55	130.90	138.98	147.73
34	69.55	76.40	83.25	86.63	**99.85**	**112.83**	**122.00**	**133.50**	141.78	151.95
35	68.98	75.85	82.70	86.10	99.43	112.52	121.80	133.50	**141.93**	152.30
36	68.15	75.15	82.00	85.37	98.80	111.98	121.35	133.25	141.87	**152.43**
40	64.24	71.10	77.85	81.27	94.65	108.03	117.55	130.15	139.43	150.52
Conversion factor: 1 S/m = 10 mS/cm (E) = 10 mmho/cm (E)										

From the above table, we can see that the highest electrical conductivity when measured at a room temperature of 20°C is 77 S/m (770 mS/cm) correspond to 30 mass percent while shifting to higher values with increasing temperature up to the maximum of 152 S/m (1520 mS/cm) for 36 mass percent at 90°C. These maxima for each temperature are depicted in the Figure 11.

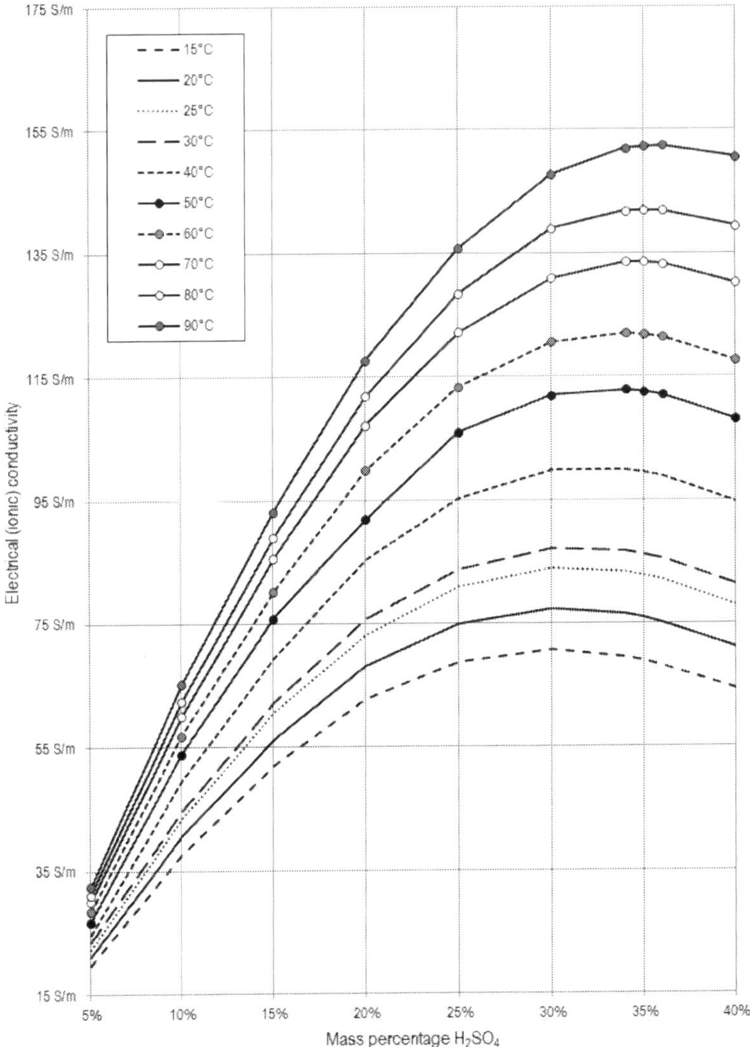

Figure 11 – Electrical conductivity vs. mass percentage of sulfuric acid

It is interesting to note that this high ionic conductivity of aqueous solutions of sulfuric acid combined with their extremely low freezing points ranging from -40°C (-40°F) for 30 wt.% and even down to -63°C (-81.4°F) for a strength of 35 wt.%, explains their widely adopted utilization, for more than a century and a half, as acidic electrolytes in the lead acid batteries invented in 1859 by the French physicist Gaston Planté.

5.9   Refractive Index

As for the mass density, the **refractive index** (n_D) of pure aqueous solutions of sulfuric acid when measured for a given wavelength (e.g., sodium D-line with λ_D = 589.3 nm) and at a fixed temperature is an important colligative property that is extremely useful for the accurate determination of the mass percentage of sulfuric acid up to a strength of 84 wt.% H_2SO_4.

Before detailing these calculations, and the determination of the refractive index of aqueous solutions of sulfuric acid, we have summarized hereafter the definitions of the four physical quantities of interest here [50].

5.9.1   Specific Refractivity and Mass Refraction

The **specific refractivity** of a substance denoted k_D) in m^3/kg, is independent of pressure and temperature and it was defined by Gladstone and Dale [51] as the ratio of **refractivity** (n_D - 1) over mass density:

$$k_D = [(n_D - 1)/\rho]$$

Similarly the **mass refraction** denoted K_D (λ_D = 589.3 nm) in m^3/kg, is used in the **Lorentz and Lorenz equation** [52] now devised from the **Clausius-Mossotti relation** used to describe dielectric materials is:

$$K_D = [(n_D^2 - 1)/(n_D^2 + 2)](1/\rho)$$

5.9.2   Molar Refractivity and Molar Refraction

The **molar retractivity**, r_D, in m^3/mol, is the product of the Gladstone's specific refraction times the molar mass of the substance:

$$r_D = Mk_D = [M(n_D - 1)/\rho]$$

The **molar refraction**, [R_D], in m^3/mol, on the other hand is the product of the mass refraction times the molar mass of the substance:

$$[R_D] = MK_D = [(n_D^2 - 1)/(n_D^2 + 2)](M/\rho)$$

---

[50] GIBB, T.R.P (1942) *Optical Methods of Chemical Analysis*. McGraw-Hill Book Company, New York, NY, pp.319-374.

[51] DALE, T. P.; and GLADSTONE, J. (1858) On the influence of temperature on the refraction of light. *Philosophical Transactions of the Royal Society London*, **148**, 887-894.

[52] LORENTZ, H. (1880) Ueber die Beziehung zwischen der Fortpflanzungs-geschwindigkeit des Lichtes und der Korperdichte. *Annalen Physik*. **245**(4)641-665.

5.9.3   Refractivity and Refraction of Sulfuric Acid Solutions

As a general rule, either the Gladstone-Dale or the Lorentz-Lorentz equations can be used for calculating the refractive properties of mixtures.

Actually, the ***molar refraction of a mixture*** of several components of molar fractions X_1, X_2, … and X_n and molar refractions $[R_1]$, $[R_2]$, … and $[R_n]$ respectively, is simply the weighted average of each constituents, and it is given by the following equation where n_D, ρ, and M refer to the refractive index, mass density, and average molar mass of the final mixture respectively:

$$[R_D] = [(n_D^2 - 1)/(n_D^2 + 2)](M/\rho) = X_1[R_1] + X_2[R_2] + \ldots + X_n[R_n]$$

with $M = \sum_i X_i M_i$

Similarly, the ***specific refractivity of a mixture*** with mass fractions w_1, w_2, … and w_n and specific refractivities k_1, k_2, … and k_n respectively is the weighted average of each constituents, and it is given by the following equation where n_D and ρ refer to the refractive index and mass density of the final mixture:

$$k_D = (n_D - 1)/\rho = w_1 k_1 + w_2 k_2 + \ldots + w_n k_n$$

The two relations above are valid over a wide range of solid, liquid or gaseous compounds with very few exceptions. Because of the

In the particular case of binary water-mixtures, the two equations become simply:

$$[R_S] = [(n_D^2 - 1)/(n_D^2 + 2)](M/\rho) = X_{H2SO4}[R_{H2SO4}] + (1 - X_{H2SO4})[R_{H2O}]$$

$$\text{with } M = X_{H2SO4}M_{H2SO4} + (1 - X_{H2SO4})M_{H2O}$$

$$= X_{H2SO4}(M_{H2SO4} - M_{H2O}) + M_{H2O}$$

$$k_S = (n_D - 1)/\rho = w_{H2SO4}k_{H2SO4} + (1 - w_{H2SO4})k_{H2O}$$

Thus based on the refractive indices and mass densities of pure water and sulfuric acid it is possible to calculate the individual refractive quantities that are summarized in Table 17.

Table 17 – Specific (molar) refractivity, refraction of water and sulfuric acid

Relations (λ_D = 589.3 nm)		Gladstone-Dale		Lorentz-Lorenz	
Substance (M_R)	Mass density and refractive index of pure substance (20°C)	k_X (cm³/g)	n_D (Mk_X) (cm³/mol)	K_X (cm³/g)	$[R_X]$ (cm³/mol)
H_2O (18.0153)	ρ = 998.204 kg/m³ n_D = 1.3330	0.33360	6.00988	0.20607	3.71233
H_2SO_4 (98.0795)	ρ = 1830.5 kg/m³ n_D = 1.4290	0.23436	22.98612	0.14084	13.81314

Example: For an aqueous solution of sulfuric acid of strength 17.0 wt.% H_2SO_4 with a tabulated mass density at 20°C of 1.1168 g/cm³, and based on an average molar mass of 20.9181874 g/mol, according to the Gladstone-Dale equations, the specific and molar refractivities are: k_S = 0.316729 cm³/g, Mk_s = 6.6259 cm³/mol respectively while with the Lorentz-Lorenz equation, the mass and molar refraction of the solution are in order: K_S = 0.19498 cm³/g, and MK_S = 4.0786 cm³/mol. From these refractive properties, the calculated refractive indices using the Gladstone-Dale and Lorentz-Lorenz equations are 1.3537 and 1.3547 respectively. The former, obtained using the Gladstone-Dale equation is an excellent agreement with the measured refractive index of n^{20}_D = 1.3538 while the latter despite still in good agreement has a greater relative error.

5.9.4   Determination of the Concentration by Refractometry

Conversely, once the refractive index and mass density of aqueous solution of sulfuric acid with an unknown strength lower than 40 wt.% are measured at the same temperature, it is then possible to calculate the mass percentage or molar fraction of sulfuric acid according to the practical equations reported in Table 18:

Table 18 – Gladstone-Dale and Lorentz-Lorenz calculations

<u>Lorentz-Lorenz (Theoretical):</u> $[R_S] = [(n_D^2-1)/(n_D^2+2)](M/\rho) = X_{H2SO4}[R_{H2SO4}] + (1-X_{H2SO4})[R_{H2O}]$ $M = X_{H2SO4}(M_{H2SO4} - M_{H2O}) + M_{H2O}$	$X_{H2SO4} = (A - B{\cdot}K_D)/(C{\cdot}K_D - D)$ $K_D = [(n_D^2 - 1)/(n_D^2 + 2)]/\rho$ $A = [R_w]$ $B = M_{H2O}$ $C = (M_{H2SO4} - M_{H2O})$ $D = [R_{H2SO4}] - [R_{H2O}]$
<u>Lorentz-Lorenz (Practical):</u> $[R_S] = [(n_D^2-1)/(n_D^2+2)](M/\rho) = 10.1012X_{H2SO4} + 3.7123$ $M = 80.0642X_{H2SO4} + 18.0153$	$X_{H2SO4} = (3.71233 - 18.0153\ K_D)/(80.0642\ K_D - 10.10081)$
<u>Gladstone-Dale (Theoretical):</u> $k_S = (n_D - 1)/\rho = (k_{H2SO4} - k_{H2O})w_{H2SO4} + k_{H2O}$	$w_{H2SO4} = (k_S - k_{H2O})/(k_{H2SO4} - k_{H2O})$
<u>Gladstone-Dale (Practical):</u> $k_S = (n_D - 1)/\rho = 0.3336 - 0.0992w_{H2SO4}$	$w_{H2SO4} = 1000(0.3336 - k_S)/99.24$

Form the above table, we can see the additional complication introduced by the average molar mass of the solution in the Lorentz-Lorenz relation and the greatest accuracy obtained with the Gladstone-Dale relation, it is easier and more accurate in practice to utilise the latter method as experienced by the author. However for sulfuric acid with strength above 40 wt.% the relative error becomes significant as exemplified below.

Example: If the refractive index of an aqueous solution of sulfuric acid when measured at 20°C with a precision Abbe refractometer using the sodium lamp as source of light is $n^{20}_D = 1.4051$ and its mass density at the same temperature is 1.4470 g/cm3, the molar fraction obtained with the practical Lorentz-Lorenz relation is 19.06 mol.% and the mass percentage obtained with the Gladstone-Dale relation is 54.05 wt.% when compared to the actual tabulated values of 20.23 mol.% and 58 wt.% respectively the relative errors of -5.8% and -6.8% respectively are two high accurate titer measurement.

Therefore, a more accurate determination of the mass percentage of sulfuric acid is obtained by performing a double interpolation from tabulated data as explained in the next paragraph.

5.9.5   Linear Interpolation Method

This method is the most accurate technique for performing the determination of the sulfuric acid strength up to 84 mass percent from the measure of its refractive index.

For that purpose, we have listed in Table 19 the refractive index measured at 20°C for the sodium D-line (λ_D = 589.3 nm), vs. the mass percentage, volume percentage, mole fraction, molarity, molality, etc. and mass density of aqueous solutions of sulfuric acid.

Table 19 – Refractive indices of sulfuric acid solutions at 20°C

Refractive index ($n^{20}{}_D$)	Mass percent (wt.%)	Volume percent (vol.%)	Mole percent (mol.%)	Mass density (kg/m³)	Conc. (g/L)	Molarity (mol/L)	Molality (mol/kg)
1.3330	0.5	0.27	0.09	1,003.90	5.02	0.051	0.0512
1.3336	1.0	0.55	0.19	1,005.10	10.05	0.102	0.1030
1.3355	2.0	1.10	0.37	1,011.80	20.24	0.206	0.2081
1.3367	3.0	1.66	0.56	1,018.40	30.55	0.312	0.3153
1.3379	4.0	2.22	0.76	1,026.40	41.06	0.419	0.4248
1.3391	5.0	2.79	0.96	1,033.20	51.66	0.527	0.537
1.3403	6.0	3.36	1.16	1,040.00	62.40	0.636	0.6508
1.3415	7.0	3.94	1.36	1,046.90	73.28	0.747	0.7674
1.3427	8.0	4.53	1.57	1,052.20	84.18	0.858	0.8866
1.3439	9.0	5.12	1.78	1,059.10	95.32	0.972	1.0084
1.3451	10.0	5.71	2.00	1,066.10	106.61	1.087	1.1329
1.3463	11.0	6.31	2.22	1,073.10	118.04	1.204	1.2602
1.3475	12.0	6.92	2.44	1,080.20	129.62	1.322	1.3903
1.3488	13.0	7.53	2.67	1,087.40	141.36	1.441	1.5235
1.3500	14.0	8.15	2.90	1,094.70	153.26	1.563	1.6598
1.3513	15.0	8.78	3.14	1,102.00	165.30	1.685	1.7993
1.3525	16.0	9.41	3.38	1,109.40	177.50	1.810	1.9421
1.3538	17.0	10.05	3.63	1,116.80	189.86	1.936	2.0883
1.3551	18.0	10.69	3.88	1,124.30	202.37	2.063	2.2381
1.3563	19.0	11.34	4.13	1,129.00	214.51	2.187	2.3916
1.3576	20.0	12.00	4.39	1,139.40	227.88	2.323	2.5490
1.3602	22.0	13.33	4.93	1,154.80	254.06	2.590	2.8757
1.3628	24.0	14.69	5.48	1,170.40	280.90	2.864	3.2197
1.3653	26.0	16.08	6.06	1,186.20	308.41	3.145	3.5823
1.3677	28.0	17.50	6.67	1,202.30	336.64	3.432	3.9650
1.3701	30.0	18.94	7.30	1,218.50	365.55	3.727	4.3696
1.3725	32.0	20.42	7.96	1,234.90	395.17	4.029	4.7980
1.3749	34.0	21.93	8.64	1,251.50	425.51	4.338	5.2524
1.3761	35.0	22.70	9.00	1,259.90	440.97	4.496	5.4901
1.3773	36.0	23.47	9.36	1,268.40	456.62	4.656	5.7351
1.3797	38.0	25.05	10.12	1,285.50	488.49	4.981	6.2490
1.3821	40.0	26.66	10.91	1,302.80	521.12	5.313	6.7972

Refractive index (n^{20}_D)	Mass percent (wt.%)	Volume percent (vol.%)	Mole percent (mol.%)	Mass density (kg/m³)	Conc. (g/L)	Molarity (mol/L)	Molality (mol/kg)
1.3846	42.0	28.31	11.74	1,320.50	554.61	5.655	7.3832
1.3870	44.0	29.99	12.61	1,338.40	588.90	6.004	8.0110
1.3883	45.0	30.85	13.06	1,347.60	606.42	6.183	8.3420
1.3895	46.0	31.72	13.53	1,356.90	624.17	6.364	8.6853
1.3920	48.0	33.48	14.50	1,375.80	660.38	6.733	9.4115
1.3946	50.0	35.29	15.52	1,395.10	697.55	7.112	10.1958
1.3971	52.0	37.14	16.60	1,414.80	735.70	7.501	11.0455
1.3998	54.0	39.03	17.74	1,435.25	775.04	7.902	11.9690
1.4011	55.0	39.99	18.33	1,445.30	794.92	8.105	12.4615
1.4024	56.0	40.97	18.95	1,455.70	815.19	8.312	12.9765
1.4051	58.0	42.96	20.23	1,477.00	856.66	8.734	14.0799
1.4077	60.0	44.99	21.60	1,498.30	898.98	9.166	15.2937
1.4102	62.0	47.08	23.06	1,520.20	942.52	9.610	16.6353
1.4128	64.0	49.22	24.62	1,542.10	986.94	10.063	18.1259
1.4140	65.0	50.32	25.44	1,553.30	1009.65	10.294	18.9351
1.4153	66.0	51.42	26.28	1,564.75	1032.74	10.530	19.7919
1.4178	68.0	53.68	28.07	1,587.40	1079.43	11.006	21.6661
1.4203	70.0	55.99	30.00	1,610.60	1127.42	11.495	23.7902
1.4229	72.0	58.37	32.08	1,633.80	1176.34	11.994	26.2178
1.4254	74.0	60.82	34.33	1,657.40	1226.48	12.505	29.0189
1.4266	75.0	62.06	35.53	1,669.20	1251.90	12.764	30.5874
1.4279	76.0	63.33	36.78	1,681.00	1277.56	13.026	32.2867
1.4304	78.0	65.91	39.44	1,704.10	1329.20	13.552	36.1488
1.4330	80.0	68.57	42.35	1,727.20	1381.76	14.088	40.7833
1.4355	82.0	71.30	45.56	1,748.25	1433.57	14.616	46.4476
1.4380	84.0	74.11	49.09	1,769.30	1486.21	15.153	53.5280
1.4369	85.0	75.55	51.00	1,778.60	1511.81	15.414	57.7763
1.4359	86.0	77.01	53.01	1,785.75	1535.75	15.658	62.6314
1.4337	88.0	80.00	57.39	1,802.20	1585.94	16.170	74.7693
1.4316	90.0	83.07	62.31	1,813.10	1631.79	16.637	91.7623
1.4295	92.0	86.25	67.87	1,824.00	1678.08	17.109	117.2518
1.4284	93.0	87.87	70.93	1,826.88	1698.99	17.323	135.4587
1.4273	94.0	89.52	74.21	1,829.75	1719.97	17.536	159.7344
1.4262	95.0	91.20	77.73	1,833.70	1742.02	17.761	193.7204
1.4252	96.0	92.90	81.51	1,835.50	1762.08	17.966	244.6995
1.4230	98.0	96.39	90.00	1,832.98	1796.32	18.315	499.5948
1.4209	100.0	100.00	100.00	1,830.45	1830.45	18.663	

From the above data, when actual refractive index n_A of an aqueous solution of sulfuric acid of unknown strength, w_A, is measured, the determination of the latter requires only to locate in the above table the closest lower and higher refractive indices, n_1 and n_2, and their respective mass percentages w_1, and w_2, and to arrange them as follows:

w_1 $\qquad\qquad\qquad\qquad$ n_1

w_A (unknown) $\qquad\quad$ n_A

w_2 $\qquad\qquad\qquad\qquad$ n_2

Thus, the mass percentage is simply given by the following relation:

$$w_A = w_1 + [(n_A - n_1)/(n_2 - n_1)](w_2 - w_1)$$

If the refractive index is measured at a different temperature, t in Celsius, it is possible to use a temperature correction equation validated for most aqueous solutions of inorganic substances:

$$n_A(20°C) = n_A(t) + 0.00045(t - 20)$$

Example: If the refractive index of an aqueous solution of sulfuric acid is 1.3698 when measured at 24°C, according to the above equation, the corrected refractive index at 20°C is 1.3716. From the tabulated data of the refractive index (20°C) vs. the mass percentage of sulfuric acid the closest values are:

$w_1 = 30$ wt.% $\qquad\qquad\qquad$ $n_1 = 1.3701$

w_A (unknown) $\qquad\qquad\quad$ $n_A = 1.3716$

$w_2 = 32$ wt.% $\qquad\qquad\qquad$ $n_2 = 1.3725$

Thus by linear interpolation, we find the unknown acid strength of 31.25 wt.% H_2SO_4.

Example: Similarly, the accuracy of above method can be improved if in in addition to measuring the refractive index, the mass density of the solution at 20°C is also measured. For the previous example, the mass density equals 1,230.0 kg/m^3, then using the Lorentz-Lorenz mass refractions instead:

$w_1 = 30$ wt.% $\qquad\qquad\qquad$ $K_1 = 0.18567$

w_A (unknown) $\qquad\qquad\quad$ $K_A = 0.18460$

$w_2 = 32$ wt.% $\qquad\qquad\qquad$ $K_2 = 0.18427$

The linear interpolation yields a sulfuric acid strength of 31.53 wt.% H_2SO_4.

In practice, when a rapid result is needed, the mass percentage can be obtained directly from the plot of the Lorentz-Lorenz mass refraction vs. the sulfuric acid strength depicted in Figure 12

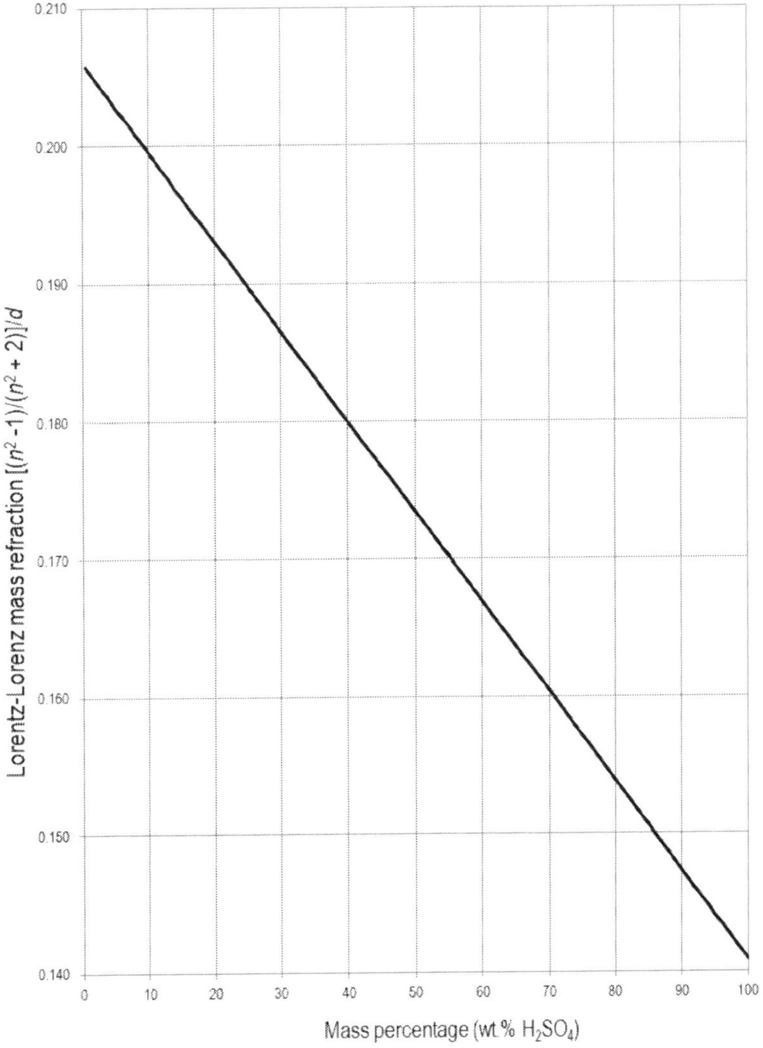

Figure 12 – Lorentz-Lorenz refraction vs. mass percentage of sulfuric acid

6 Corrosion Resistant Materials

Because the sulfuric acid digestion is usually performed inside large containment vessels also called "digesters" capable to withstand both the high temperature and elevate corrosiveness of sulfuric acid without contaminating the products by releasing deleterious metallic impurities the selection of proper construction materials is of paramount importance.

Actually, sulfuric acid exhibits a strong corrosiveness towards various construction materials both metallic and non-metallic. Moreover, its corrosiveness and the corrosion mechanisms varies widely as a function of the acid concentration, the operating temperature, the presence of deleterious impurities such as reducing or oxidizing species, and in a lesser extent the pressure.

Additionally, because in most sulfation methods, the ubiquitous presence of suspended solids within the strong and hot sulfuric acid, promote the abrasion and erosion of materials under significant fluid velocities and thus represent one of the most challenging conditions the chemical industry is experiencing.

In this particular situation, a very limited number of heat and corrosion resistant materials are available commercially that are capable to withstand such drastic conditions with expected service life of several years in some instance.

Collected corrosion data from manufacturer provides iso-corrosion curves for various metals and alloys with a corrosion rate established at 127 μm/year or 5 mils per year (mpy).

Hereafter, we will briefly describe the major classes of materials used industrially with their advantages and drawbacks along with their limitation. For a very deep coverage on the subject, the reader is directed to consult the comprehensive set of industrial and corrosion data published by the *Deutsche Gesellschaft für chemisches Apparatewesen* (DECHEMA) [53].

---

[53] BEHRENS, D.; ESKERMANN, R.; and KREYSA, G. (eds.) (1991) *DECHEMA Corrosion Handbook, Sulfuric Acid, Volume 8, Eighth Edition.* Wiley-VCH, Weinheim, Germany.

Another important point to consider when combatting corrosion in concentrated sulfuric acid is the risk of corrosion from 'self-dilution' of sulfuric acid. Actually, as sulfuric acid exhibits a strong affinity for water, absorbing water from its surroundings and hence diluting it. The result can be that sulfuric acid thought to be concentrated say above 93 wt.% H_2SO_4, can actually corrodes the containment materials if water has been picked up. This can occur especially in open tanks where moisture from the air dilutes the acid and results in corrosion around the liquid-line. As heat is generated when sulfuric acid is diluted hotter conditions can be present locally, which can add to the risk of attack in the more dilute acid.

6.1   Metals and Alloys

6.1.1   Lead and its Alloys

Because ***pure lead*** forms a protective and impervious layer of insoluble lead sulfate ($PbSO_4$) at its surface exposed to sulfate containing medium, it was historically used in the lead chamber process to withstand strong sulfuric acid at concentration up to 93 wt.% H_2SO_4 [54]. However, this protective film is easily removed by erosion and abrasion and thus handling suspended solids often result in accelerated degradation.

Table 20 – Solubility of lead sulfate in sulfuric acid

Mass percentage of sulfuric acid (H_2SO_4)	Solubility $PbSO_4$ (mg/L)	
	20°C	50°C
10 wt.%	5.6	10.7
20 wt.%	4.4	7.3
30 wt.%	2.9	4.9
40 wt.%	1.9	4.1
96 wt.%	0.5	3.0

Moreover, the inherent softness of lead and its alloys and the creeping preclude its use for structural parts or when dimensional stability is required. For that purpose, increased strength and hardness, lead is usually alloyed with antimony, tin, silver or calcium to impart greater mechanical properties. The presence of impurities such as chloride is extremely detrimental. Nowadays, lead and its alloys are still used sometimes as protective membrane between inert bricks and the carbon steel shell for industrial digester despite being phased out in favor of fluoropolymers.

---

[54] KUHN, A.T. (ed.)(1979) *The Electrochemistry of Lead.* Academic Press, London, UK, pp. 365-381.

6.1.2   Cast Irons and High Silicon Cast Iron

The direct consequence of the relatively low solubilities of iron (II) sulfate explains why *cast irons* are to some extent satisfactory for handling concentrated sulfuric acid at ambient temperature for concentration ranging from 65 wt.% H_2SO_4 up to 100 wt.% H_2SO_4. As a general rule, increased levels of carbon and silicon are beneficial for imparting a greater corrosion resistance.

As a matter of fact, the French metallurgist Jouve discovered in 1908 that high silicon cast irons containing 14 wt.% up to 16 wt.% silicon offer exceptional resistance to all concentration up to 100 wt.% H_2SO_4 due the impervious film of silicon dioxide (silica) that forms onto the surface [55]. Since its first production in the United States in 1912, the well-known commercial alloy named ***Duriron***® (Trademark of *The Duriron Company, Inc.*, Dayton, Ohio) consisting of Fe-14.75Si-0.35Mn-0.8C exhibits a remarkable corrosion resistance towards sulfuric acid for a wide range of concentrations from ambient temperature up to the boiling point. Selected corrosion rates for Duriron® in sulfuric acid solutions at 80°C and at the boiling point are reported in Table 21. However, high silicon cast irons are not suitable for handling oleum as they corrode significantly. It is worth to note on the other hand, that traces of Fe(III) improves the passivation and thus the corrosion resistance of Duriron® in dilute sulfuric acid solutions..

The selected physical properties of high silicon cast iron with 14.5 w.% Si are as follows: a mass density of 7,050 kg/m³, a thermal expansion coefficient from 0°C to 100°C of 12.5 x 10^{-6} K^{-1}, an electrical resistivity of 63.3 $\mu\Omega$.cm, and a thermal conductivity of 52 W.m^{-1}.K^{-1}.

The mechanical properties are an ultimate tensile strength of 100-140 MPa, a compressive strength of 655-690 MPa, with a Brinell hardness of 500 and with a Shore scleroscope hardness of 49-51. The latter demonstrates the excellent abrasion resistance that is suitable for handling suspensions of abrasive solids [56].

---

[55] RILEY, R.V. (1992) Silicon iron - the acid resisting alloy. *Journal of Historical Metallurgy Society (JHMS)* **26**, 56-60.

[56] BORGES VILLARINO DE CASTRO, D.; DE SOUSA MALAFAIA, A.M., ANGELONI, M.; SGARBI ROSSINO, L.; and MALUF,O. (2009) Behavior of high-silicon cast iron under wear and corrosion. *20th International Congress of Mechanical Engineering*, November 15-20, 2009, Gramado, RS, Brazil.

Table 21 – Corrosion rate of high silicon cast iron in sulfuric acid

Mass percentage of sulfuric acid (wt.% H_2SO_4)	Corrosion rate (mpy)	
	80°C	Boiling point
10	30	80
30	38	70
50	40	80
60	14	60
77.7	nil	nil
96	nil	nil
Oleum 30%	130	n.a.

Therefore high silicon cast irons castings are often used when the harshest conditions are encountered such as pipes, valves, fittings, pans, mixers, compressors, and pumps. Their only drawback is that they can only be produced as castings with a high melting range of 1246°C to 1300°C and a casting contraction allowance of 21 mm per meter.

6.1.3   Mild and Carbon Steels

As for cast irons, *mild steel*, and *carbon steels* are relatively resistant to corrosion when exposed to concentrated sulfuric acid at ambient temperature because of the low solubility of the impervious film of iron (II) sulfate that forms onto the exposed surface. That's why mild steel was selected for constructing tank cars for the transportation of the concentrated sulfuric acid by railroad or truck at ambient temperature. When a greater corrosion resistance is required or the temperature increased above ambient carbon steels can be protected with a coating of enamel or small autoclaves are even entirely glass lined. The latter option is however prone to catastrophic failure in case of mechanical impact or thermal shock.

6.1.4   Austenitic Stainless Steels

The direct consequence of the extremely low solubilities of iron (II), and nickel (II) sulfates in concentrated sulfuric acid explains why *austenitic stainless steels* exhibits a good corrosion resistance at ambient temperature with corrosion rates below one mil per year (25.4 µm/year) in concentrated sulfuric acid above 93 wt.% H_2SO_4 for type AISI 304 (Fe-18.5Cr-10Ni-1.5Mn) [57]. Moreover, stainless steels are more suitable than carbon steels for handling high flow rates of concentrated acid (90-98 wt.% H_2SO_4). The passive layer onto stainless steels is more stable than the ferrous sulfate layer formed on carbon steel in turbulent flow

---

[57] KIEFER, G.C.; and RENSHAW, W.G. (1950) The behavior of the chromium nickel stainless steels in sulfuric acid. *Corrosion*, 6(8)235-244

conditions. Flow rates can become a problem as the active/passive region of concentration and temperature is approached.

However, the upper temperature threshold established at 40°C restricts its use to piping, fluid handling equipment, and storage tanks. The types AISI 316 (Fe-17Cr-13Ni-2.25Mo-1.8Mn) and AISI 317 (Fe-19Cr-14Ni-3.25Mo-2Mn) due to their additional 2 wt.% molybdenum content extend the corrosion resistance up to 90 wt.% H_2SO_4 for the same temperature range. The resistance can however be increased by maintaining artificially the stainless steel into its passive state by anodic protection [58]. In this case, the technique consists to impose a DC current between the part to protect usually a tank and suitable cathode materials immune to hydrogen embrittlement. It is worth to mentioned heat resistant austenitic stainless steel grade AISI 310 (Fe-25Cr-20Ni-2Mn) is only resistant at high concentration as its corrosion resistance decreases markedly below 96 wt.% H_2SO_4. Commercially, the proprietary grades **Saramet**® (Trademark of *Chemetics Inc.*), **Sandvick SX**® (Trademark of *Sandvick AB*), and **ZeCor**® (Trademark of *MECS, Inc.*) are austenitic stainless steels containing 5 wt.% silicon to impart improved corrosion resistance at high flow velocities for hot concentrated sulfuric acid with corrosion rates less than 25.4 µm per year (1 mil per year).

6.1.5 Duplex Stainless Steels

Duplex stainless steels constitute a new class of corrosion resistant alloys are specifically alloyed and processed to develop a microstructure with equal amounts of ferrite and austenite phases. Since they have high content of chromium and molybdenum they exhibit a good corrosion resistance double to that of annealed austenitic stainless steels with regard to pitting, crevice corrosion, sulfide stress corrosion, and chloride stress corrosion environments. The corrosion behavior of duplex stainless steels in various concentrations of sulfuric acid is especially suitable when the sulfuric acid is contaminated by chloride anions. For instance, the grade **SAF 2205** (Trademark of *Sandvick AB*) roughly Fe-Cr25-Ni7-Mo4 shows much better resistance than austenitic stainless steel grade 316L and has similar resistance to Alloy 904L (45Fe-25Ni-21Cr-4.7Mo-2.2Cu-2Mn) in such medium. The latter was specifically developed for sulfuric acid use across the whole concentration range up to 35°C. Moreover, the commercial grades of **superduplex stainless steels** such as **SAF 2507** (Trademark of *Sandvik AB*) and **Zeron®100** (Trademark of *Weir Materials Limited*) are particularly designed to withstand hot concentrated sulfuric acid the latter due to tungsten and copper alloying elements..

6.1.6 Nickel-based Alloys

Hastelloy® C-276 (Trademark of *International Nickel Company, Inc.*) (54Ni-16Mo-15.5Cr-5Fe-2.5Co) has a wide application range in sulfuric acid

---

[58] JUCKNIEWICZ, R.; POMPOWSKI, T.; and WALASZKOWSKI, J. (1966) Anodic protection of austenitic stainless steels. Corrosion Science, **6**, 25-31.

---

environments [59]. At ambient temperature, its corrosion rate, at all concentrations, is below 4 mils per year (100 μm per year). Because of its elevate chromium content, it is particularly suitable for handling sulfuric acid containing oxidizing impurities (e.g., Fe^{3+}, Cu^{2+}) especially when compared to ***Hastelloy® B2*** (67Ni-28Mo-1Cr-2Fe) (Trademark of *International Nickel Company, Inc.*) that resists corrosion towards concentrated sulfuric acid much better under reducing conditions.

6.1.7   Refractory Metals

6.1.7.1   Zirconium

Pure ***zirconium*** metal because of its valve action property is protected by a thin and impervious passivation layer of ordered cubic zirconia (ZrO_2). This passive film ensures an excellent corrosion resistance towards sulfuric acid up to its boiling point and up to an upper concentration of 60 wt.% H_2SO_4 (i.e., *bp* = 144°C). Above these upper limits, the significant corrosion precludes the utilization of zirconium without affecting its dimensional stability. For instance the corrosion rate reaches 20 mils per year (500 μm per year) in boiling 75 wt.% H_2SO_4. The corrosion of zirconium, by contrast with titanium, is even strongly affected by the presence of even traces of oxidizing species such as ferric or cupric cations at the part per million (ppm vol.) levels. Commercially, the most resistant grade is the ***Zircadyne® 702*** (*Trademark of ATI-Wah Chang*). For more information for the utilization of Zirdadyne®, it is strongly advised to consult the exhaustive collection of corrosion data from the supplier [60].

6.1.7.2   Tantalum

Tantalum metal despite its prohibitive price exhibits exceptional corrosion resistance similar to some extent to pure platinum and quartz. The domain covers the entire range of temperature from room temperature up to the boiling point and for acid concentration up to 98 wt.% H_2SO_4. Above the latter concentration at high temperature when the oxidizing properties of concentrated sulfuric acid becomes significant tantalum is corroded readily and it must not be used. This is even more severe with pyrosulfuric acid ($H_2S_2O_7$) and oleums. This is why immersed tantalum heaters or tube and shell heat exchangers are often used in the last stage of evaporators to produce 96 wt.% H_2SO_4.

6.1.8   Platinum Group Metals

Historically because of its peculiar chemical inertness, pure ***platinum*** foil were used once as loose lining to clad cast iron vessels used to evaporate the

[59] COLLECTIVE (1983) *The Corrosion Resistance of Nickel-Containing Alloys in Sulfuric Acid and Related Compounds.* The International Nickel Company (INCO), Inc., Suffern, NY, pp. 35-36.

[60] ANONYMOUS (1997) *Zircadyne® - Zirconium Corrosion Data. Technical Data Sheet,* Wah Chang, Albany, OR.

concentrated sulfuric acid originating from the lead acid chamber. Nowadays, except for laboratory use such as crucibles, platinum and its alloy are no more used industrially to handle concentrated sulfuric acid.

6.1.9   Corrosion Resistance Chart for Metals and Alloys

For convenience, we have categorized in Table 22 and Table 23, the corrosion resistance of selected commercial metal and alloys and high silica glass for comparison towards solutions of sulfuric acid of various strength as a function of the mass percentage of sulfuric acid ranging from 20 wt.% up to 100 wt.% H_2SO_4 and the temperature ranging from 20°C up to 100°C.

The three letter-code was established as follow: **A** excellent with a corrosion rate below 2 mils per year (50.8 μm/year), **B** good with a corrosion rate below 5 mils per year (127 μm/year), and finally **C** not suitable due to severe corrosion.

However, in practice for selecting the proper material, we strongly recommend utilizing the specific corrosion data provided by each metal or alloy manufacturer along with the numerous isocorrosion plots established from experimental corrosion data that are provided in the scientific and technical literature on the subject.

Finally, the final selection of the suitable corrosion resistant material shall always be based on extensive laboratory testing using standardized coupons the actual sulfuric acid solution or acidic solution under the same operating conditions such as temperature, pressure, volume flow rate, solids concentration, oxidation-reduction potential (ORP), and aeration, etc.

Table 22 – Corrosion selection chart for sulfuric acid strength from 20% to 70%

Concentration H₂SO₄	20 w/o				30 w/o					40 w/o					50 w/o				60 w/o				70 w/o			
Temperature (°C)	50	60	80	100	20	40	60	80	100	20	40	60	90	100	20	40	70	100	20	40	70	100	20	40	70	100
Carbon steel	C	C	C	C	C	C	C	C	C	C	C	C	C	C	C	C	C	C	C	C	C	C	C	C	C	C
Stainless 420	C	C	C	C	C	C	C	C	C	C	C	C	C	C	C	C	C	C	C	C	C	C	C	C	C	C
Stainless 304/304L	C	C	C	C	C	C	C	C	C	C	C	C	C	C	C	C	C	C	C	C	C	C	C	C	C	C
Stainless 316/316L	B	C	C	C	B	C	C	C	C	C	C	C	C	C	C	C	C	C	C	C	C	C	C	C	C	C
Stainless 317/317L	B	B	C	C	B	B	C	C	C	C	C	C	C	C	C	C	C	C	C	C	C	C	C	C	C	C
17Cr14Ni4Mo	B	B	C	C	B	B	C	C	C	C	C	C	C	C	C	C	C	C	C	C	C	C	C	C	C	C
904L	A	A	B	C	A	A	B	C	C	A	A	B	C	C	A	A	C	C	A	B	B	C	A	B	B	C
Alloy 20 Cb																										
Noram SX™																										
Zeron® 100			B	C			B	C	C			B	C	C		B	C	C		B	C	C		B	C	C
Sanicro® 28	A	A	C	C	A	A	B	C	C	A	A	B	C	C	A	A	B	C	A	A	B	C	A	A	B	C
254 SMO	A	A	C	C	A	A	B	C	C		B			C	A	B		C	A	B		C	A	B		C
654 SMO	A	A	A	C						A	A	A		C	A	A		C	A	B		C	A	B		C
SAF 2205	A	B	C	C	A	B	C	C		C	C	C	C	C	C	C	C	C	C	C	C	C	B			C
SAF 2304	C	C	C	C	C	C	C	C		C	C	C	C	C	C			C				C				C
SAF 2507	A	A				A	B			A			C	C			C	C				C	A	C	C	C
SAF 2707 HD	C	C	A	C			C				A	A			C	C	C		C	C	C		B			
SAF 2906	C	C	A	C			C				A	C			C	C	C		C	C	C		A			
SAF 3207 HD	C	C	A	C			C				A	A			A	C	C		C	C	C		A			
Titanium Gr. 2	C	C	C	C	C	C	C	C	C	C	C	C	C	C	C	C	C	C	C	C	C	C	C	C	C	C
Hastelloy® B2	C	C	C	C	B	B	B	C	C	B	B	B	C	C	B	A	A	B	A	A	A	B	A	A	A	B
Hastelloy® C-276	A	A	A	C	A	A	A	A	C	A	A	A	A	C	A	A	A	C	A	A	A	C	A			C
Zircadyne® 702	A	A	A	A	A	A	A	A	A	A	A	A	A	A	A	A	A	A	A	B	B	C	B	B	B	C
Niobium	A	A	A	C	A	A	A	A	C	A	A	A	A	C	A	A	A	C	A	A	A	A	A	A	A	A
Tantalum	A	A	A	A	A	A	A	A	A	A	A	A	A	A	A	A	A	A	A	A	A	A	A	A	A	A
Vycor® glass	A	A	A	A	A	A	A	A	A	A	A	A	A	A	A	A	A	A	A	A	A	A	A	A	A	A

Table 23 – Corrosion selection chart for sulfuric acid strength from 80% to 98%

Concentration H_2SO_4	Temperature (°C)	Carbon steel	Stainless 420	Stainless 304/304L	Stainless 316/316L	Stainless 317/317L	17Cr14Ni4Mo	904L	Alloy 20 Cb	Noram SX™	Zeron® 100	Sanicro® 28	254 SMO	654 SMO	SAF 2205	SAF 2304	SAF 2507	SAF 2707 HD	SAF 2906	SAF 3207 HD	Titanium Gr 2	Hastelloy® B2	Hastelloy® C-276	Zircadyne® 702	Niobium	Tantalum	Vycor® glass
98 w/o	80	C	C	C	C	C	C	C		A	B	B	C	B	B	B	B				C			C	C	A	A
	50	C	C	C	A	B	B	B		A	B	A	A	B	B	A	A				C			C	C	A	A
	40	B	B	A	A	A	A	B		A	B	A				A		A			C			C	C	A	A
	30	B	B	A	A	A	A	A		A	B	A				A		A			C			C	C	A	A
96 w/o	100	C								A	B										C	B	C	C	C	A	A
	50	C	C	B	B	B	B	B		A	B	B		C		A	B				C			C	C	A	A
	40	C	C	A	A	B	B	B		A	B	A				B		A			C			C	C	A	A
	30	B	B	A	A	A	A	A		A	B	A				B		A			C			C	C	A	A
	20	A	A	A	A	A	A	A		A	B	A	B	A	A	B	A				C			C	C	A	A
94 w/o	100	C								A	C										C	B	C	C	C	A	A
	50	C	C	B	B	B	B	B		A	B	A		C			B				C			C	C	A	A
	40	C	C	B	A	B	B	B		A	B	A				A					C			C	C	A	A
	30	C	B	A	A	A	A	A		A	B	A				A					C			C	C	A	A
	20	A	A	A	A	A	A	A		A	A	A				A				A	C	A		C	C	A	A
90 w/o	100	C	C	C	C	C	C	C			C	C	C	C	C	C					C	B	C	C	C	A	A
	70	C	C	C	C	C	C	C	A		B		C				C	C	C	C	C	A	A	C	C	A	A
	40	C	C	C	B	B	B	B	A	B	A	B	C	B	B	A					C	A	B	C	C	A	A
	30	B	B	A	A	B	B	A	A	B	A				B	B	A				C	A	B	C	C	A	A
	20	A	A	A	A	A	A	A	A	A	A	B	B	B			A				C	A	B	C	C	A	A
85 w/o	100	C	C	C	C	C	C	C			C	C	C	C	C	C					C	B	C	C	C	A	A
	50	C	C	C	C	C	C	B		A	C	A									C	A		C	C	A	A
	40	C	C	B	B	B	B	B		A	B	A					B	B	B	B	C	A	B	C	C	A	A
	30	B	B	B	B	B	B	A		A	B	A						B	B		C	A	B	C	C	A	A
	20	A	B	B	B	B	B	A		A	A	A	A					B	B	B	C	A	B	C	C	A	A
80 w/o	100	C	C	C	C	C	C	C			C	C	C	C	C	C					C	B	C	C	C	A	A
	60	C	C	C	C	C	C	C		C	B	C		C			C				C	A	B	C	C	A	A
	40	C	C	C	C	C	C	B	A	B	B	B		C			C	C	C	C	C	A	B	C	C	A	A
	20	C	C	C	C	C	B	C	A		A	B	C				C				C	A	B	C	C	A	A

6.2 Glass and Vycor

Borosilicate glass (Pyrex®) (Trademark of _Corning Glass Works_) is resistant towards sulfuric acid from all the concentrations and up to the boiling point. **Vycor®** (Trademark of *Corning Glass Works*) with 96 wt.% SiO_2 is even more resistant and used for wetted parts operating at high temperature.

6.3 Ceramics and Refractories

6.3.1 Quartz

Quartz in the form of **vitreous silica** and **fused silica** is much more chemically resistant than borosilicate glass and less prone to thermal shocks due to its lower expansion coefficient (10^{-6} K^{-1}) and thus it is used for wetted parts operating at high temperature without showing any sign of corrosion after several years of continuous operation. This is one of the main reasons why commercial evaporators and sulfuric acid recovery plants are all built using piping made entirely of glass and fused silica such as those manufactured worldwide by the leading company *QVF-DeDietrich Process Systems GmbH*, Mainz, Germany.

6.3.2 Alumina and High Alumina Bricks

Pure **aluminium oxide (alumina)** (Al_2O_3) in the form of corundum shows excellent resistance to 93 wt.% H_2SO_4 up to 95°C with no noticeable corrosion. Dense electro fused white and brown alumina (99 wt.% Al_2O_3) exhibits chemical inertness towards dilute aqueous solution of sulfuric acid [61] but the chemical resistance is lower for strong and hot sulfuric acid 93 wt.% H_2SO_4 up to 95°C with 12 mils per year (304.9 µm per year).

6.3.3 Zirconia

Electro fused **zirconia** exhibits a good corrosion rate for strong sulfuric acid at 93 wt.% H_2SO_4, however, the corrosion rate becomes greater than one mil per year (25.4 µm per year) above 95°C.

6.3.4 Silicon Carbide

Sintered **silicon carbide**, SiC, sold under the tradenames **Carborundum®** and **Hexoloy®** (both Trademarks of *Saint-Gobain Abrasives, Inc.*) or is a relatively dense (3,100 kg/m³) and non-porous ceramic which is extremely corrosion resistant towards 98 wt.% sulfuric acid at 100°C (8 µm/year) and it is used often as filler in acid-proof mortars [62]. Some polymorphs phases even exhibit outstanding corrosion resistance towards mixture of sulfuric, hydrofluoric,

[61] CURKOVIC, L.; FUDURIC, M.; and KURAJICA, S. (2008) Chemical stability of alumina ceramics in sulphuric acid. *Materials Testing*, **50**(6)336-340.

[62] JEON, I.K.; QUDOOS, A.; JAKHRANI, S.H.; KIM, H.G.; and RYOU, J.-S. (2020) Investigation of sulfuric acid attack upon cement mortars containing silicon carbide powder. *Powder Technology*, **359**, 181-189.

and nitric acids [63]. Besides, silicon carbide possesses outstanding physical properties such as a high thermal conductivity (130 $W.m^{-1}.K^{-1}$) as well as extremely high compression (2,900 MPa), tensile and flexural (460 MPa) strengths making it suitable as filler for acid proof mortars and immersed electric heaters.

6.3.5   Cast Basalt

Volcanic igneous rocks were originally used in France during the 19th century, to build equipment to handle concentrated sulfuric acid for the lead chamber process. The two fine grained igneous rocks suited for that particular application were ***trachy-andesite*** also called *Pierre de Volvic* and in a lesser extent ***basalt*** both quarried in the Massif Central due to the availability of large naturally occurring slabs offering a smooth finish with a low porosity after polishing and lapping.

Nowadays, naturally occurring basalt rock is melted at 1280°C and cast into vitreous preforms. After subsequent heat treatment of the preforms to re-crystallize and then produce a more resistant artificial materials exhibiting a denser non porous form the pipes of cast basalt or cast basalt lined steel are produced. Due to the excellent chemical inertness to acids and alkalis and its abrasion resistance it is sold under the tradenames **Abresist®** from *Kalenborn* in the USA and **Basmarite®** from the company *Eutit* in the Czech Republic.

6.4   Polymers and Elastomers

Several fluorinated thermoplastics of exceptional corrosion resistance are extensively used as protective lining materials for instance: PVC, CPVC, TFE, PVDF, PTFE, and PFA.

6.4.1   Polyvinyl Chloride (PVC)

Polyvinyl chloride, PVC (Type I) also known as **Corzan®** (Trademark of *Lubrizol Advanced Materials, Inc.*), is an affordable construction material that chemically withstands all concentrations of sulfuric acid up to 80 wt.% H_2SO_4 below a maximum working temperature of 60°C (140°F). Because PVC has poor temperature resistance, it should not be used as construction material piping and components that can be exposed to transient high temperature resulting from the the heat released during dilution of the acid.

---

[63] KRASOTKINA, N.I.; YAKOVLEVA, V.S.; VORONIN, N.I.; and SHMITT-FOGELEVICH, S.P. (1968) Stability of silicon carbide to hydrofluoric, nitric, and sulfuric acids. *Refractories*, **9**, 723-726.

6.4.2   Chlorinated Polyvinyl Chloride (CPVC)

Chlorinated Polyvinyl Chloride, CPVC (Type I) is an affordable construction material that chemically withstand all concentrations of sulfuric acid up to 80 wt.% H_2SO_4 but with a higher maximum service temperature of 93°C (200°F) compared to PVC.

6.4.3   Polyvinylediene Fluoride (PVDF)

Polyvinylidene fluoride, PVDF, also known as **Kynar®** (Trademark of *Arkema, Inc.*) exhibits an excellent corrosion resistance for all acid concentrations up to 93 wt.% H_2SO_4. Above this strength it not recommended as it will exhibit some degradation over time. Its higher working temperature of up to 107°C (225°F) makes an excellent choice for handling warm concentrated acid safely.

6.4.4   Polytetrafluoroethylene (PTFE)

Polytetrafluoroethylene, PTFE, also known as **Teflon®** (Trademark of *The Chemours Company FC, LLC*), along with other perfluorinated polymers such as perfluoroalkoxy, PFA, is outstanding as it is the only material that can withstand all concentrations of sulfuric acid at the temperatures up to a maximum working temperature of about 204°C (400°F). This material is however expensive and mechanically soft, thus pipes of vessels made of PTFE-lined steel are often used instead combining the outstanding chemical resistance of PTFE and the inherent stiffness of carbon steel.

6.4.5   Chlorotrifluoroethylene (CTFE)

Chlorotrifluoroethylene, CTFE, also known as **Halar®** (Trademark of *Solvay Solexis Spa.*) can withstand concentrated sulfuric acid but exhibits a much lower maximum temperature 121°C (250°F) than PTFE, and because it is more expensive, and can be fabricated only with limited shapes and pipe fitting available. Finally it discolors to a brown or yellow hue when exposed with concentrated sulfuric acid.

6.4.6   Epoxy Resin Fiberglass

Epoxy resin fiberglass exhibits good chemical resistance to sulfuric acid with strength up to 70 wt.% H_2SO_4 and temperature up to about 42°C (180°F). However, care must be taken to ensure that strong concentrated acid is not spilled on the fiberglass, as this will degrade the epoxy and cause failure.

6.4.7   Compatibility Chart for Polymers and Elastomers

For convenience, we have categorized in Table 24, the compatibility at ambient temperature of selected commercial polymers and elastomers towards sulfuric acid of various strength.

The four letter-codes were established as follow: **A**: excellent corrosion resistance with no sign of deterioration or discoloration; **B**: good with only slight corrosion, or discoloration; **C**: fair not recommended for continuous use; and finally **D**: severe deterioration not recommended for any use.

In practice for selecting the proper polymer, we strongly recommend utilizing the compatibility data provided by plastic manufacturers.

Finally, the final selection of the suitable polymer or elastomer shall always be based on extensive laboratory testing using standardized protocols using the actual sulfuric acid solution under the same operating conditions such as temperature, pressure, volume flow rate, solids concentration, oxidation-reduction potential (ORP), and aeration, etc.

Table 24 – Compatibility chart for polymers and elastomers

Polymer or Elastomer	Trademarks	H_2SO_4 below 10 wt%	H_2SO_4 10 to 75 wt%	H_2SO_4 75 to 98 wt%	Cold H_2SO_4 98 to 100 wt%	Hot H_2SO_4 98 to 100 wt%
Acetal	Delrin®	D - Poor	D - Poor	N/A	N/A	N/A
Acrylonitrile butadiene styrene (ABS)		B - Good	B - Good	N/A	N/A	N/A
Blend of PPO, PPE and PS	Noryl®	A - Excellent	A - Excellent	A - Excellent	A - Excellent	D - Poor
Buna N (Nitrile)		A - Excellent	B - Good	C - Fair	D - Poor	D - Poor
Chlorinated polyvinyl chloride (CPVC)	Corzan®	A - Excellent	A - Excellent	C - Fair	D - Poor	D - Poor
Chlorosulfonated polyethylene (CSPE)	Hypalon®	A - Excellent	B - Good	C - Fair	C - Fair	D - Poor
Epoxy resin		A - Excellent	A - Excellent	C - Fair	D - Poor	D - Poor
Ethylene propylene diene monomer (EPDM)		A - Excellent	B - Good	B - Good	C - Fair	D - Poor
Fluorocarbon	Viton®	A - Excellent	A - Excellent	A - Excellent	B - Good	A - Excellent
Low-density polyethylene (LDPE)		A - Excellent	A - Excellent	B - Good	C - Fair	D - Poor
Natural rubber (NR)		A - Excellent	C - Fair	D - Poor	D - Poor	D - Poor
Perfluoroelastomer	Kalrez®	A - Excellent	A - Excellent	A - Excellent	A - Excellent	N/A
Polyamides	Nylon®	C - Fair	D - Poor	D - Poor	D - Poor	D - Poor
Polycarbonate (PC)	Lexan®	A - Excellent	B - Good	D - Poor	N/A	D - Poor
Polychloroprene (Neoprene)		B - Good	B - Good	D - Poor	D - Poor	D - Poor
Polychlorotrifluoroethylene (PCTFE)	Kel-F®	A - Excellent	A - Excellent	A - Excellent	A - Excellent	A - Excellent
Polyetherether Ketone (PEEK)		A - Excellent	D - Poor	D - Poor	D - Poor	D - Poor
Polyphenylene sulfide (PPS)	Ryton®	A - Excellent	A - Excellent	A - Excellent	A - Excellent	D - Poor
Polypropylene (PP)		A - Excellent	A - Excellent	C - Fair	A - Excellent	D - Poor
Polytetrafluoroethylene (PTFE)		A - Excellent	A - Excellent	A - Excellent	A - Excellent	A - Excellent
Polyurethane (PU)		D - Poor	D - Poor	D - Poor	D - Poor	D - Poor
Polyvinyl chloride (PVC)	Corzan®	A - Excellent	A - Excellent	A - Excellent	D - Poor	D - Poor
Polyvinylidene fluoride (PVDF)	Kynar®	A - Excellent	A - Excellent	A - Excellent	A - Excellent	C - Fair
Silicone	Tygon®	B - Good	N/A	N/A	D - Poor	D - Poor

6.5 Composite Approach

Another efficient solution extensively used industrially consists to construct digesters made of a several layers of materials each of them having a particular protecting or structural role.

In this approach, a thick welded carbon steel shell acting as a structural component is lined internally with a first impervious layer made of plastics, elastomers or formerly sheets of corroding lead metal acting as a protective membrane.

Then a second lining made of acid-proof refractory bricks offer a protection from the heat and the abrasion of solids.

The acid-proof refractories are made of high silica bricks that are assembled with an acid resistant mortar made of a mixture silica flour and potassium silicate.

Sometimes, a foamed-glass intermediate layer is installed between the bricks and the membrane. Examples of such composite walls are depicted in Figure 13.

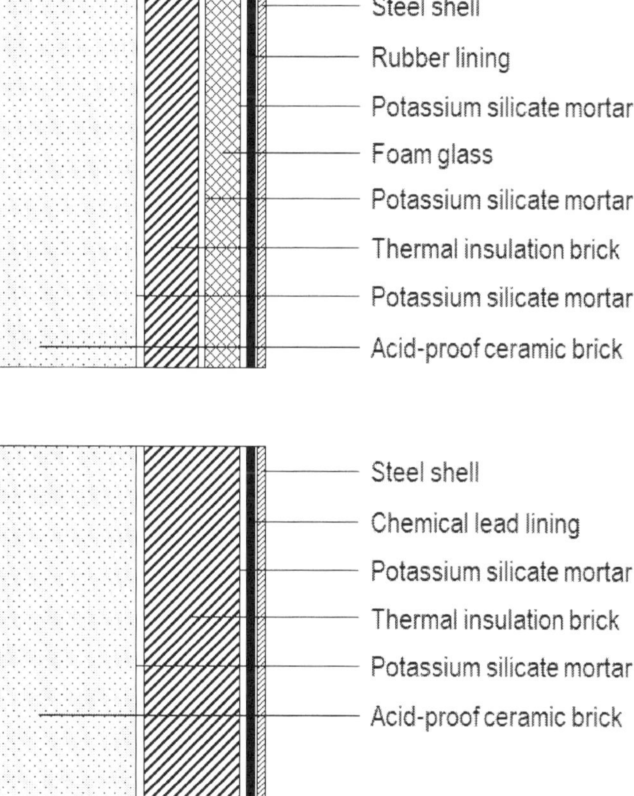

Figure 13 – Examples of digesters composite walls

7 Sulfuric Acid Numbers and Mass Ratios

7.1　Theoretical Sulfuric Acid Number

The theoretical mass of sulfuric acid required to perform a sulfation reaction is based only on the stoichiometry of the sulfation reaction.

Therefore, the dimensionless ***sulfuric acid-to-solid mass ratio*** also called the ***theoretical sulfuric acid number***" or ***stoichiometric acid number*** or simply "***acid number***", is defined as the mass ratio of pure sulfuric acid (100 wt.% H_2SO_4) to the mass of dry solid, S, denoted either by (H_2SO_4-to-S), [H_2SO_4-to-S], [H_2SO_4:S] or by the acronym *TAN*.

The stoichiometric sulfuric acid-to-solid mass ratio is calculated from the actual chemical composition of the dry solid, expressed as metal oxides, assuming a stoichiometric consumption of sulfuric acid by each reactive metal oxide to be sulfated and that they form the most stable metal sulfate at the given operating temperature. Note that some oxides such as silicon dioxide (silica) do not react with sulfuric acid and then do not take part in the consumption of sulfuric acid for these calculations.

For instance, we have listed in Table 25 the individual theoretical acid-to-solid mass ratios of selected metal oxides assuming a stoichiometric consumption of 100% sulfuric acid per unit mass of single oxide (kg/tonne).

Table 25 – Stoichiometric consumption of 100% sulfuric acid per unit mass

Sulfation chemical reaction schemes	Theoretical consumption of 100% sulfuric acid per unit mass of solid (kg/tonne)
$BeO + H_2SO_4 = BeSO_4 + H_2O(g)$	3,920
$Li_2O(s) + H_2SO_4 = Li_2SO_4(s) + H_2O(g)$	3,282
$Al_2O_3 + 3H_2SO_4 = Al_2(SO_4)_3 + 3H_2O(g)$	2,886
$MgO + H_2SO_4 = MgSO_4 + H_2O(g)$	2,433
$P_2O_5 + 3H_2SO_4 = 2H_3PO_4 + 3SO_3(g)$	2,073
$Ti_2O_3 + 3H_2SO_4 = Ti_2(SO_4)_3 + 3H_2O(g)$	2,047
$V_2O_3 + 3H_2SO_4 = V_2(SO_4)_3 + 3H_2O(g)$	1,963
$Cr_2O_3 + 3H_2SO_4 = Cr_2(SO_4)_3 + 3H_2O(g)$	1,936
$Fe_2O_3 + 3H_2SO_4 = Fe_2(SO_4)_3 + 3H_2O(g)$	1,843
$CaO + H_2SO_4 = CaSO_4 + H_2O(g)$	1,749
$Fe_3O_4 + 4H_2SO_4 = Fe_2(SO_4)_3 + FeSO_4 + 4H_2O(g)$	1,694
$Na_2O(s) + H_2SO_4 = Na_2SO_4(s) + H_2O(g)$	1,582
$MnO + H_2SO_4 = MnSO_4 + H_2O(g)$	1,383
$FeO + H_2SO_4 = FeSO_4 + H_2O(g)$	1,365
$NiO + H_2SO_4 = NiSO_4 + H_2O(g)$	1,313
$ZnO + H_2SO_4 = ZnSO_4 + H_2O(g)$	1,205
$CaF_2 + H_2SO_4 = CaSO_4 + 2HF(g)$	1,256
$TiO_2 + H_2SO_4 = TiOSO_4 + H_2O(g)$	1,228
$K_2O(s) + H_2SO_4 = K_2SO_4(s) + H_2O(g)$	1,041
$La_2O_3 + 3H_2SO_4 = La_2(SO_4)_3 + 3H_2O(g)$	903
$ThO_2 + 2H_2SO_4 = Th(SO_4)_2 + 2H_2O(g)$	743
$V_2O_5 + H_2SO_4 = (VO_2)_2SO_4 + 3H_2O(g)$	539

Therefore, based on the actual chemical composition of a material to be sulfated, the overall acid-to-solid mass number, $[H_2SO_4\text{-to-}S]_{total}$, is the sum of all the weighted averages of the individual acid numbers of each single reactive metal oxide $[H_2SO_4\text{-to-}S]_k$:

$$[H_2SO_4\text{-to-}S]_{total} = \sum_k [H_2SO_4\text{-to-}S]_k$$

For example the sulfation of a dry solid made of a spodumene concentrate with the following chemical composition (in wt.%): 68 SiO_2, 24 Al_2O_3, 6.0 Li_2O, and 0.76 Fe_2O_3 will require a stoichiometric sulfuric-to-solid mass ratio

[H_2SO_4-to-S] equal to 930 kg (100 wt.% H_2SO_4) per tonne of dry concentrate if we assume that all the lithium and aluminum values are sulfated.

It is important to mention, that often in practice, a mass of concentrated sulfuric acid, A, with its strength reported as a mass percentage w_A is used instead of 100% sulfuric acid per unit mass of dry solid, S. Therefore the liquid-to-solid mass ratio denoted either by (A-to-S), [A-to-S], or [A:S] is related to the theoretical mass ratio by the straightforward equation:

$$[H_2SO_4\text{-to-S}] = w_A\,[A\text{-to-S}]$$

In the previous example, if "winter acid" (93 wt.% H_2SO_4) is used the [A-to-S] becomes 1,000 kg per tonne to achieve the same sulfation.

7.2   Sulfuric Acid Number per Single Oxide

In some instances, the sulfuric acid number is calculated as the dimensionless mass ratio of initial 100% sulfuric acid to a selected oxide, compound or product of interest to ensure a certain free acidity once the sulfation cake is dissolved into water.

This is the especially the case in the titanium pigment industry, in the phosphate fertilizer industry, in the lithium industry or during the sulfation of bauxite and other alumina feedstocks where the sulfuric acid numbers are reported per unit mass of pure titanium dioxide [H_2SO_4-to-TiO_2], pure phosphorus pentoxide [H_2SO_4-to-P_2O_5], pure lithium oxide [H_2SO_4-to- Li_2O] or pure alumina [H_2SO_4-to- Al_2O_3] respectively.

7.3   Practical Sulfuric Acid Number and Excess Acid

In common industrial practice, an excess of sulfuric acid is often required to compensate for its loss by evaporation, by decomposition, by absorption by inert or unreacted solids or in order to maintain a suitable pulp density with targeted rheological properties, a proper wettability, and suitable fluidity allowing the full dissolution of the products.

The excess of sulfuric acid usually ranges from 10 to 25 mass percent but in some cases can be up to 200 mass percent or more (e.g., 2000%). The latter figures are specifically utilized when the sulfation reaction is expected to be strongly exothermic and the sensible heat of the diluted suspension of the solids is required to prevent a catastrophic thermal runaway by absorbing the excess energy released during the sulfation reaction.

This is particularly true when performing the sulfuric acid digestion of feedstock's and raw materials containing high mass percentage of CaO and/or

---

MgO such as basic metallurgical slags (e.g., BOF-slags, LD-slags) (see Sections 8.1 and 10.3).

8 Thermochemistry of Sulfation Reactions

One significant advantage for performing the sulfation roasting and/or the sulfuric acid digestion of complex ores and concentrates containing various metal oxides is the fact that most sulfation reactions are strongly exothermic and despite the large evolution of heat and related temperature elevation of the reacting mixture (reactants and products), the elevate boiling point of concentrated sulfuric acid allows to operate efficiently with minimum losses.

Moreover, it is important when implementing a future industrial process and from an occupational safety standpoint to assess the total amount of heat released due to the exothermic sulfation reaction for the following reasons in order: (1) to avoid a catastrophic thermal runaway when processing larger batches; (2) to maintain the targeted operating temperature inside a safe zone especially regarding the operating limit of the containment materials used industrially; and (3) to evaluate the minimum amount of heat required for the sulfation process to be self-sustainable and autogenous from an energy standpoint.

8.1 Specific Enthalpy and Gibbs Enthalpy of Sulfation

For the sake of clarity, we will consider here only the most often encountered chemical reactions of sulfation that are listed in Table 1. As a general rule, the sulfation of a single metal oxide is described by the following chemical reaction:

$$M_xO_y(s) + yH_2SO_4(l) = M_x(SO_4)_y(s) + yH_2O(g)$$

$$\textbf{Table 26} - \text{Selected chemical reactions of sulfation at } 180°C$$

$Li_2O(s)$	$+ H_2SO_4$	$=$	$Li_2SO_4(s)$	$+ H_2O(g)$
$Na_2O(s)$	$+ H_2SO_4$	$=$	$Na_2SO_4(s)$	$+ H_2O(g)$
$K_2O(s)$	$+ H_2SO_4$	$=$	$K_2SO_4(s)$	$+ H_2O(g)$
$BeO(s)$	$+ H_2SO_4$	$=$	$BeSO_4(s)$	$+ H_2O(g)$
$MgO(s)$	$+ H_2SO_4$	$=$	$MgSO_4(s)$	$+ H_2O(g)$
$CaF_2(s)$	$+ H_2SO_4$	$=$	$CaSO_4(s)$	$+ 2HF(g)$
$CaO(s)$	$+ H_2SO_4$	$=$	$CaSO_4(s)$	$+ H_2O(g)$
$Ca(OH)_2(s)$	$+ H_2SO_4$	$=$	$CaSO_4(s)$	$+ 2H_2O(g)$
$FeO(s)$	$+ H_2SO_4$	$=$	$FeSO_4(s)$	$+ H_2O(g)$
$MnO(s)$	$+ H_2SO_4$	$=$	$MnSO_4(s)$	$+ H_2O(g)$
$NiO(s)$	$+ H_2SO_4$	$=$	$NiSO_4(s)$	$+ H_2O(g)$
$ZnO(s)$	$+ H_2SO_4$	$=$	$ZnSO_4(s)$	$+ H_2O(g)$
$Al_2O_3(s)$	$+ 3H_2SO_4$	$=$	$Al_2(SO_4)_3(s)$	$+ 3H_2O(g)$
$Cr_2O_3(s)$	$+ 3H_2SO_4$	$=$	$Cr_2(SO_4)_3(s)$	$+ 3H_2O(g)$
$Fe_2O_3(s)$	$+ 3H_2SO_4$	$=$	$Fe_2(SO_4)_3(s)$	$+ 3H_2O(g)$
$La_2O_3(s)$	$+ 3H_2SO_4$	$=$	$La_2(SO_4)_3(s)$	$+ 3H_2O(g)$
$Ti_2O_3(s)$	$+ 3H_2SO_4$	$=$	$Ti_2(SO_4)_3(s)$	$+ 3H_2O(g)$
$V_2O_3(s)$	$+ 3H_2SO_4$	$=$	$V_2(SO_4)_3(s)$	$+ 3H_2O(g)$
$Fe_3O_4(s)$	$+ 4H_2SO_4$	$=$	$Fe_2(SO_4)_3(s)$	$+ 4H_2O(g) + FeSO_4(s)$
$FeTiO_3(s)$	$+ 2H_2SO_4$	$=$	$FeSO_4(s)$	$+ 2H_2O(g) + TiOSO_4(s)$
$TiO_2(s)$	$+ H_2SO_4$	$=$	$TiOSO_4(s)$	$+ H_2O(g)$
$ThO_2(s)$	$+ 2H_2SO_4$	$=$	$Th(SO_4)_2(s)$	$+ 2H_2O(g)$
$V_2O_4(s)$	$+ 2H_2SO_4$	$=$	$2VOSO_4(s)$	$+ 2H_2O(g)$
$V_2O_5(s)$	H_2SO_4	$=$	$(VO_2)_2SO_4(s)$	$+ H_2O(g)$

The equation for the temperature dependence of each thermodynamic function (i.e., enthalpy, entropy, Gibbs free enthalpy) for the sulfation of a single metal oxide, which is reactive towards sulfuric acid, can be calculated using tabulated thermodynamic properties available in the literature [64, 65, 66, 67].

For instance, the thermodynamic properties of the metal oxides and few compounds along with the corresponding metal sulfates produced are reported in Table 27.

[64] LIDE, D.R. (ed.) (2001) *Handbook of Chemistry and Physics. 81st Edition.* CRC Press Bocca Raton, FL; pp. 5-85 to 5-88.

[65] WICKS, C.E.; and BLOCK, F.E. (1963) *Thermodynamic Properties of 65 Elements, Their Oxides, Halides, Carbides, and Nitrides.* Bulletin No. 605, U.S. Bureau of Mines, United States Government Printing Office, Washington D.C.

[66] BARIN, I.; and KNACKE, O. (1973) *Thermochemical Properties of Inorganic Substances.* Springer-Verlag, Heidelberg, Germany.

[67] KUBASCHEVSKI, O.; BALBCOK, C.B.; and SPENCER, P.J. (1993) *Materials Thermochemistry, Sixth Edition.* Pergamon Press, Oxford, UK.

Table 27 – Thermodynamic properties of selected substances

Chemical substance (physical state, mineral phase)	Std. molar enthalpy ΔH^o J/mol	Std. molar entropy S^0 J/(mol.K)	a_k J/(mol.K)	b_k J/(mol.K^2)	c_k J.K/mol
$Al_2(SO_4)_3$ (solid)	-3,435,064	239.32	366.31	0.0626	-1.116E+07
Al_2O_3 (alumina)	-1,673,600	51.04	114.77	0.0128	-3.544E+06
BeO (solid)	-598,730	14.14	21.22	0.0551	-8.678E+05
$BeSO_4$ (solid)	-1,200,808	77.97	71.78	0.0997	-1.378E+06
CaF_2 (fluorite)	-1,214,615	68.95	59.83	0.0305	-1.966E+05
$Ca(OH)_2$ (portlandite)	-986,211	83.39	105.29	0.0119	-1.897E+06
CaO (solid)	-634,294	39.75	49.62	0.0045	-6.945E+05
$CaSO_4$ (anhydrite)	-1,432,602	106.69	70.21	0.0987	
$CaSO_4.0.5H_2O$ (solid)	-1,575,150	130.54	107.93	0.0987	
$CaSO_4.2H_2O$ (gypsum)	-2,021,123	193.97	221.08	0.0987	
$Cr_2(SO_4)_3$ (solid)	-2,931,298	258.80	358.07	0.0795	-8.980E+6
Cr_2O_3 (eskolaite)	-1,129,680	81.17	119.37	0.0092	-1.565E+06
$Fe_2(SO_4)_3$ (solid)	-2,582,992	307.52	309.62	0.1116	-6.422E-04
Fe_2O_3 (hematite)	-825,503	87.45	98.28	0.0778	-1.485E+06
Fe_3O_4 (magnetite)	-1,118,383	146.44	86.27	0.2089	
FeO (wustite)	-272,044	60.75	50.80	0.0086	-3.310E+05
$FeSO_4$ (solid)	-928,848	120.96	122.01	0.0378	-2.929E-04
HF (gas)	-271,100	173.78	29.13		
H_2O (liquid)	-285,830	69.50	75.35		
H_2O (vapor)	-242,463	188.72	30.00	0.0107	3.347E+04
H_2SO_4 (liquid)	-813,989	156.90	156.90	0.0284	-2.346E+06
H_3PO_4 (solid)	-1,284,400	110.54	15.48	0.3037	-9.682E+05
K_2O (solid)	-363,171	94.14	72.17	0.0418	
K_2SO_4 (solid)	-1,433,857	175.73	120.37	0.0996	-1.782E-04
$La_2(SO_4)_3$ (solid)	-3,941,300		280.00		

Chemical substance (physical state, mineral phase)	Std. molar enthalpy ΔH^0 J/mol	Std. molar entropy S^0 J/(mol.K)	a_κ J/(mol.K)	b_κ J/(mol.K^2)	c_κ J.K/mol
La$_2$O$_3$ (solid)	-1,793,262	128.03	120.75	0.0129	-1.372E+06
Li$_2$O (solid)	-598,730	37.89	69.58	0.0179	-1.904E+06
Li$_2$SO$_4$ (solid)	-1,321,700	115.10	117.60		
MgO (periclase)	-601,241	26.94	48.98	0.003142184	-1.144+06
MgSO$_4$ (solid)	-1,284,906	91.63	106.44	0.04627504	-2.190E04
MnO (solid)	-384,928	59.83	46.48	0.0081	-3.682E+05
MnSO$_4$ (solid)	-1,065,246	112.13	122.42	0.0373	2.946E-04
Na$_2$O (solid)	-417,982	75.06	63.96	0.0439	-8.134E+05
Na$_2$SO$_4$ (solid)	-1,387,205	149.62	82.32	0.1544	
NiO (bunsenite)	-240,580	37.99	-20.88	0.1572	1.628E+06
NiSO$_4$ (solid)	-889,100	92.00	125.94	0.0278	-3.264E-04
P$_4$O$_{10}$ (solid)	-2,984,029	228.86	70.04	0.4519	
SO$_2$ (gas)	-296,813	237.20	43.43	0.0106	-5.941E+05
SO$_3$ (gas)	-395,765	256.65	57.15	0.0273	-1.291E+06
Th(SO$_4$)$_2$ (solid)	-2,541,362	148.11	104.60	0.2310	-
ThO$_2$ (solid)	-1,226,749	65.27	69.66	0.0089	-9.372E+05
Ti$_2$(SO$_4$)$_3$ (solid)	-3,387,000	254.50	272.30	-	-
Ti$_2$O$_3$ (tistarite)	-1,520,842	78.78	152.60	0.0000	-5.004E+06
TiO$_2$ (rutile)	-944,747	50.33	62.86	0.01136	-0.996E+06
TiOSO$_4$ (solid)	-1,637,000	95.50	112.50	-	-
V$_2$(SO$_4$)$_3$ (solid)	-3,240,508	-408.78	-	-	-
V$_2$(SO$_4$)$_3$ (solid)	-2,336,668	-374.30	-	-	-
V$_2$O$_3$ (karelianite)	-1,225,912	98.32	122.80	0.0199	-2.268E+06
V$_2$O$_4$ (lenoblite)	-1,435,112	102.93	62.59	-	-
V$_2$O$_5$ (solid)	-1,557,703	130.96	194.72	-0.0163	-5.531E+06
VOSO$_4$ (solid)	-1,309,200	108.00	-	-	-

Chemical substance (physical state, mineral phase)	Std. molar enthalpy ΔH° J/mol	Std. molar entropy S^0 J/(mol.K)	a_k J/(mol.K)	b_k J/(mol.K²)	c_k J.K/mol
$(VO_2)_2SO_4$ (solid)	-2,208,940	-66.10	-	-	-
ZnO (solid)	-348,109	43.51	48.99	0.0051	-9.121E-05
$ZnSO_4$ (solid)	-978,638	30.60	91.63	0.0761	-
Data from BARIN, I.; and KNACKE, O. (1973) *Thermochemical Properties of Inorganic Substances.* Springer-Verlag, Heidelberg.					

The temperature dependence of the **molar heat capacity** of a single chemical substance k denoted C_{pk}, in $J.K^{-1}.mol^{-1}$, can be described by the following empirical equation:

$$C_{pk}(T) = a_k + b_k T + c_k/T^2$$

Where the three **empirical heat capacity coefficients** a_k in $J.K^{-1}.mol^{-1}$, b_k in $J.K^{-2}.mol^{-1}$, and c_k in $J.K.mol^{-1}$, are determined from calorimetric experiments.

On the other hand, for a given single chemical species k, at a given temperature T in K, its **standard molar enthalpy**, denoted $H^0_k(T)$, in $J.mol^{-1}$, and **standard molar entropy**, denoted $S^0_k(T)$, in $J.K^{-1}.mol^{-1}$, can be calculated using the two following equations:

$$H^0_k(T) = H^0_k(T_0) + \int C_{pk} dT + \sum \Delta H_{tr}$$

$$S^0_k(T) = S^0_k(T_0) + \int [C_{pk}/T] dT + \sum \Delta S_{tr}$$

Where $H^0_k(T_0)$ and $S^0_k(T_0)$ are the standard enthalpy and entropy at the reference temperature (T_0 = 298.15K) with $\sum \Delta H_{tr}$ and $\sum \Delta S_{tr}$ are the sums of the **latent molar enthalpies** and **latent molar entropies** of all the transitions (e.g., phase change, fusion, vaporization, and sublimation) occurring within the selected temperature range.

Hence, after a straightforward mathematical integration, we obtain:

$$H^0_k(T) = H^0_k(T_0) + a_k(T - T_0) + 0.5 b_k(T^2 - T_0^2) + c_k(1/T_0 - 1/T) + \sum \Delta H_{tr}$$

$$S^0_k(T) = S^0_k(T_0) + a_k(\ln T - \ln T_0) + b_k(T - T_0) + 0.5 c_k(1/T_0^2 - 1/T^2) + \sum \Delta S_{tr}$$

Regrouping all the constant terms together in bracket followed by the variables, we obtain the following two concise equations for the temperature dependence of the standard molar enthalpy and entropy:

$$H^0_k(T) = [H^0_k(T_0) - a_k T_0 - 0.5b_k T_0^2 + c_k/T_0 + \sum \Delta H_{tr}] + a_k T + 0.5b_k T^2 - c_k/T$$

$$S^0_k(T) = [S^0_k(T_0) - a_k \ln T_0 - b_k T_0 + 0.5c_k/T_0^2 + \sum \Delta S_{tr}] + a_k \ln T + b_k T - 0.5c_k/T^2$$

Afterwards, based on the definition of the molar ***Gibbs molar enthalpy*** below:

$$G^0_{(T)} = H^0(T) - TS^0(T)$$

It is possible to establish the temperature dependence of the standard Gibbs molar enthalpy of each single component as:

$$G^0_k(T) = H^0_k(T_0) - TS^0_k(T_0) + a_k(T - T_0) + 0.5b_k(T^2 - T_0^2) + c_k(1/T_0 - 1/T) - T[a_k(\ln T - \ln T_0) + b_k(T - T_0) + 0.5c_k(1/T_0^2 - 1/T^2)]$$

As previously, regrouping all the constant terms together in bracket followed by the variables, we obtain the following condensed equation for the temperature dependence of the molar Gibbs enthalpy:

$$G^0_k(T) = [H^0_k(T_0) - a_k T_0 - 0.5b_k T_0^2 + c_k/T_0 + \Delta H_{tr}] + [a_k(1 + \ln T_0) + b_k T_0 - S^0_k(T_0) - 0.5c_k/T_0^2]T - 0.5b_k T^2 - 0.5c_k/T - a_k T \ln T$$

Based on the three above equations, for $H^0_k(T)$, $S^0_k(T)$, and $G^0_k(T)$, we can establish the temperature dependence of the standard molar enthalpy, $\Delta_r H^0(T)$, standard molar entropy, $\Delta_r S^0(T)$, and molar Gibbs enthalpy, $\Delta_r G^0(T)$, for a given chemical sulfation reaction:

$$\Delta_r G^0(T) = \Delta_r H^0(T) - T\Delta_r S^0(T)$$

with $\quad \Delta_r H^0(T) = \sum_k v_k H^0_k(T) \quad$ and $\quad \Delta S^0(T) = \sum_k v_k S^0_k(T)$

Where v_k are the dimensionless stoichiometric coefficients. By convention, these are arbitrarily taken positive for products ($v_k > 0$) and negative ($v_k < 0$) for reactants.

At this point, it is worth to introducing the differential quantities:

$$\Delta a = \sum_k v_k a_k$$

$$\Delta b = \sum_k v_k b_k$$

$$\Delta c = \sum_k v_k c_k$$

Based on all the above equations, it is possible to establish the temperature dependence of the standard molar enthalpy and molar Gibbs enthalpy for the sulfation reaction:

$$\Delta_r H^0(T) = [\Delta_r H^0{}_k(T_0) - \Delta a T_0 - 0.5\Delta b T_0{}^2 + \Delta c/T_0 + \Sigma_k \, v_k \, \Delta H_{tr}] + \Delta a T + 0.5\Delta b T^2 - \Delta c/T$$

$$\Delta_r G^0(T) = [\Delta H^0(T_0) - \Delta a T_0 - 0.5\Delta b T_0{}^2 + \Delta c/T_0 + \Sigma_k \, v_k \, \Delta H_{tr}] + [\Delta a(1 + \ln T_0) + \Delta b T_0 - \Delta S^0(T_0) - 0.5\Delta c/T_0{}^2]T - 0.5\Delta b T^2 - 0.5\Delta c/T - \Delta a T \ln T$$

We have summarized the two above equations along with their more concise forms utilising their related empirical coefficients A, B, C, D, and E and A^*, B^*, C^*, and D^* in Table 28.

Table 28 – Temperature dependence of molar and Gibbs sulfation enthalpies

Molar Gibbs enthalpy of sulfation	$\Delta_r G^0(T) = [\Delta_r H^0(T_0) - \Delta a T_0 - 0.5\Delta b T_0{}^2 + \Delta c/T_0 + \Sigma_k \, v_k \, \Delta H_{tr}] + [\Delta a(1 + \ln T_0) + \Delta b T_0 - \Delta_r S^0(T_0) - 0.5\Delta c/T_0{}^2]T - 0.5\Delta b T^2 - 0.5\Delta c/T - \Delta a T \ln T$ $\Delta_r G^0(T) = A + BT + CT^2 + D/T + ET\ln T$ with $A = [\Delta_r H^0(T_0) - \Delta a T_0 - 0.5\Delta b T_0{}^2 + \Delta c/T_0 + \Sigma_k \, v_k \, \Delta H_{tr}]$ $B = [\Delta a(1 + \ln T_0) + \Delta b T_0 - \Delta_r S^0(T_0) - 0.5\Delta c/T_0{}^2]$ $C = -0.5\Delta b$ $D = -0.5\Delta c$ $E = -\Delta a$
Molar enthalpy of sulfation	$\Delta_r H^0(T) = [\Delta_r H^0(T_0) - \Delta a T_0 - 0.5\Delta b T_0{}^2 + \Delta c/T_0 + \Sigma_k \, v_k \, \Delta H_{tr}] + \Delta a T + 0.5\Delta b T^2 - \Delta c/T$ $\Delta_r H^0(T) = A^* + B^*T + C^*T^2 + D^*/T$ with $A^* = [\Delta_r H^0(T_0) - \Delta a T_0 - 0.5\Delta b T_0{}^2 + \Delta c/T_0 + \Sigma_k \, v_k \, \Delta H_{tr}]$ $B^* = \Delta a$ $C^* = 0.5\Delta b$ $D^* = -\Delta c$

Therefore, based on the two empirical equations above, we can calculate the molar enthalpy and molar Gibbs energy for each selected sulfation reaction listed in Table 26 for a temperature ranging from 298.15K (25°C) and 453.15K (180°C).

For that purpose, we have used the thermodynamic properties (e.g., standard molar enthalpy, standard molar entropy, and empirical coefficients of the specific heat capacity) of each reactant and product reported in Table 27.

Thus we obtained the practical equations with their respective empirical coefficients that are reported altogether in Table 29. Moreover, for convenience, we have also calculated the molar enthalpy (kJ/mol) and the specific enthalpies (kJ/kg) for the sulfation reaction at an absolute temperature of 453.15K (180°C).

Note that we have arbitrarily chosen this temperature interval as it is often encountered during sulfuric acid digestion conditions and also because all the reactants and products do not exhibit transition changes within that range.

If a higher upper temperature is utilized, then, it will be important to take into account all the related latent molar enthalpies and entropies along with the new set of empirical coefficients for the specific heat capacity.

Table 29 – Coefficients for the molar and Gibbs enthalpies of sulfation reactions

| Metal oxide and relative molar mass | Thermodynamic function | Empirical coefficients $\Delta_r G(T) = A + BT + CT^2 + D/T + E\,T\ln T$ $\Delta_r H(T) = A^* + B^*T + C^*T^2 + D^*/T$ | | | | | $\Delta_R H_{453K}$ | $\Delta_R h_{453K}$ |
		A, A^*	B, B^*	C, C^*	D, D^*	E	kJ/mol	kJ/kg
Li_2O	$\Delta G(T)$	-111,976	-672.042	1.779E-02	-2.142E+06	78.88064	-199	-6,658
29.8814	$\Delta H(T)$	-19,103	-332.682	-3.550E-02	8.943E+06		-197	**-6,589**
Na_2O	$\Delta G(T)$	-358,078	-838.719	-4.639E-02	-1.597E+06	110.79232	-444	-7,166
61.9789	$\Delta H(T)$	-265,205	-364.594	2.868E-02	7.853E+06		-442	**-7,129**
K_2O	$\Delta G(T)$	-469,604	-641.780	-2.001E-02	-1.173E+06	78.70104	-549	-5,828
94.1960	$\Delta H(T)$	-376,619	-332.502	2.297E-03	7.039E+06		-542	**-5,758**
BeO	$\Delta G(T)$	-2,718	-609.386	-1.346E-02	-9.349E+05	76.33290	-72	-2,884
25.0116	$\Delta H(T)$	90,155	-330.134	-4.257E-03	6.529E+06		-75	**-2,988**
CaF_2	$\Delta G(T)$	86,874	-821.928	-1.99284E-02	-1.272E+06	88.25768	-48	-613
78.0748	$\Delta H(T)$	190,014	-372.925	-8.498E-03	7.236E+06		3	**+42**
CaO	$\Delta G(T)$	-188,172	-805.303	-3.825E-02	-1.537E+06	106.31544	-270	-4,809
56.0774	$\Delta H(T)$	-95,299	-360.117	2.054E-02	7.734E+06		-271	**-4,839**
$Ca(OH)_2$	$\Delta G(T)$	-67,066	-1128.305	-3.990E-02	-2.155E+06	131.98846	-225	-3,043
74.0927	$\Delta H(T)$	35,115	-415.789	1.683E-02	8.936E+06		-170	**-2,289**
FeO	$\Delta G(T)$	-60,092	-476.857	-5.745E-03	-1.355E+06	55.69285	-126	-1,754
71.8444	$\Delta H(T)$	32,781	-309.494	-1.197E-02	7.370E+06		-126	**-1,756**
MgO	$\Delta G(T)$	-80,745	-573.867	-1.271E-02	-1.762E+06	69.44603	-155	-3,841
40.3044	$\Delta H(T)$	12,128	-323.247	-5.006E-03	8.183E+06		-153	**-3807**
MnO	$\Delta G(T)$	-84,892	-437.468	-5.745E-03	-1.374E+06	50.96112	-146	-2,060
70.9375	$\Delta H(T)$	7,981	-304.763	-1.197E-02	7.407E+06		-149	**-2,099**
NiO	$\Delta G(T)$	-73,870	-0.528	7.356E-02	-3.761E+05	-19.91744	-115	-1,540
74.6928	$\Delta H(T)$	19,003	-233.884	-9.128E-02	5.412E+06		-118	**-1,575**
ZnO	$\Delta G(T)$	-28,267	-580.770	-2.666E-02	-1.190E+06	84.26576	-66	-811
81.3894	$\Delta H(T)$	64,606	-338.067	8.950E-03	7.039E+06		-102	**-1,257**
Al_2O_3	$\Delta G(T)$	-9,836	-1147.104	1.678E-03	2.397E+05	129.16008	-171	-1,675

Metal oxide and relative molar mass	Thermodynamic function	Empirical coefficients $\Delta_r G(T) = A + BT + CT^2 + D/T + E\,T\ln T$ $\Delta_r H(T) = A^* + B^*T + C^*T^2 + D^*/T$					$\Delta_R H_{453K}$	$\Delta_R b_{453K}$
		A, A^*	B, B^*	C, C^*	D, D^*	E	kJ/mol	kJ/kg
101.9613	$\Delta H(T)$	-10,173	-129.160	-1.678E-03	-5.799E+05		-68	**-665**
Cr_2O_3	$\Delta G(T)$	-46,389	-1217.495	-8.575E-03	1.378E+05	142.00168	-206	-1,355
151.9904	$\Delta H(T)$	-46,725	-142.002	8.575E-03	-3.760E+05		-108	**-714**
Fe_2O_3	$\Delta G(T)$	37,374	-1504.160	9.689E-03	-4.312E+06	169.36432	-182	-1,142
159.6882	$\Delta H(T)$	37,037	-169.364	-9.689E-03	8.524E+06		-61	**-379**
La_2O_3	$\Delta G(T)$	-335,949	-1518.191	3.302E-02	-4.256E+06	221.45240	-413	-2,716
151.9904	$\Delta H(T)$	-336,286	-221.452	-3.302E-02	8.412E+06		-462	**-3,040**
Ti_2O_3	$\Delta G(T)$	-30,670	-2103.402	2.657E-02	-6.072E+06	261.00101	-268	-1,867
143.7322	$\Delta H(T)$	-31,007	-261.001	-2.657E-02	1.204E+07		-181	**-1,261**
V_2O_3	$\Delta G(T)$	-115,099	-3035.311	3.653E-02	-4.704E+06	503.50256	-98	-653
149.8812	$\Delta H(T)$	-115,436	-503.503	-3.653E-02	9.307E+06		-372	**-2,480**
Fe_3O_4	$\Delta G(T)$	-21,259	-1588.342	6.518E-02	-4.760E+06	162.23859	-288	-1,246
231.5326	$\Delta H(T)$	-30,142	-157.348	-7.524E-02	7.039E+06		-132	**-572**
$FeTiO_3$	$\Delta G(T)$	-537,939	-1037.730	3.772E-02	-3.123E+06	117.57360	-681	-4,492
151.7102	$\Delta H(T)$	-458,129	-356.974	-4.657E-02	8.524E+06		-648	**-4,273**
TiO_2	$\Delta G(T)$	-90,930	-521.266	-6.215E-03	-1.688E+06	63.81855	-155	-1,944
79.8658	$\Delta H(T)$	1,943	-317.620	-1.150E-02	8.035E+06		-162	**-2,029**
ThO_2	$\Delta G(T)$	-95,493	-1588.756	-9.331E-02	-2.848E+06	218.86504	-234	-887
264.0369	$\Delta H(T)$	-49,225	-345.766	8.445E-02	7.976E+06		-206	**-781**
V_2O_4	$\Delta G(T)$	71,636	-2333.128	1.772E-02	-2.380E+06	316.39408	-110	-665
165.8806	$\Delta H(T)$	117,904	-443.295	-2.657E-02	7.039E+06		-104	**-627**
P_4O_{10}	$\Delta G(T)$	511,259	-4662.996	-3.783E-01	-1.229E+06	606.63649	-1	-3
283.8890	$\Delta H(T)$	488,356	-353.822	-7.568E-02	6.071E+06		299	**1054**
V_2O_5	$\Delta G(T)$	42,778	-2033.783	6.987E-04	-3.956E+06	321.62408	4	22
181.8800	$\Delta H(T)$	135,651	-575.426	-1.841E-02	1.257E+07		-157	**-861**

8.2 Dehydration of Hydrated Metal Sulfates

Hydrated metal sulfates, often encountered during sulfation reactions, can lose their water molecules progressively yielding intermediate hydrates and finally the anhydrous metal sulfate.

Table 30 – Dehydration temperature of hydrated metal sulfates

Dehydration reactions	Dehydration temperature or range	Melting point
$FeSO_4.7H_2O = FeSO_4.4H_2O + 3H_2O(g)$	26°C - 96°C	-
$FeSO_4.4H_2O = FeSO_4.H_2O + 3H_2O(g)$	100°C - 140°C	-
$3FeSO_4.H_2O + 1.5O_2 = Fe_2(SO_4)_3 + 0.5Fe_2O_3 + 3H_2O(g)$	492°C - 560°C	decompose
$Fe_2(SO_4)_3 + 3H_2O(g)$	492°C - 560°C	decompose
$MnSO_4.H_2O = MnSO_4 + H_2O(g)$	250°C	700°C
$MgSO_4.7H_2O = MgSO_4.6H_2O + H_2O(g)$	36°C – 60°C	-
$MgSO_4.6H_2O = MgSO_4.5H_2O + H_2O(g)$	103°C	-
$MgSO_4.5H_2O = MgSO_4.3H_2O + H_2O(g)$	108°C	-
$MgSO_4.3H_2O = MgSO_4.2H_2O + H_2O(g)$	127°C	-
$MgSO_4.2H_2O = MgSO_4.H_2O + H_2O(g)$	149°C	-
$MgSO_4.H_2O = MgSO_4 + H_2O(g)$	178°C – 285°C	1100°C
$BeSO_4.4H_2O = BeSO_4 + 4H_2O(g)$	400°C	-
$CaSO_4.2H_2O = CaSO_4.0.5H_2O + 1.5H_2O(g)$	110°C	-
$CaSO_4.0.5H_2O = CaSO_4 + 0.5H_2O(g)$	186°C	
$Li_2SO_4.H_2O = Li_2SO_4 + H_2O(g)$	72°C - 160°C	877°C
$Na_2SO_4.10H_2O = Na_2SO_4.4.75H_2O + 5.25H_2O(g)$	72°C	-
$Na_2SO_4.5.25H_2O = Na_2SO_4 + 5.25H_2O(g)$	90°C	880°C
$Cr_2(SO_4)_3.17H_2O(g) = Cr_2(SO_4)_3.15H_2O + 2H_2O(g)$	80°C	-
$Cr_2(SO_4)_3.15H_2O(g) = Cr_2(SO_4)_3.6H_2O + 9H_2O(g)$	180°C	-
$Cr_2(SO_4)_3.6H_2O(g) = Cr_2(SO_4)_3 + 6H_2O(g)$	500°C	decompose

8.3 Thermal Decomposition of Metal Sulfates

Once the dehydration of metal sulfates is completed, it is important to understand the possible further decomposition of the anhydrous metal sulfates formed if the decomposition temperature is reached

This is especially true during techniques such as sulfation baking and sulfation roasting with operating temperatures greater than 300°C.

From the data reported in the scientific and technical literature [68, 69] regarding the decomposition temperature of various anhydrous metal sulfates, it is then of paramount importance for industrial processes to establish a maximum temperature threshold in order to avoid the thermal decomposition of the anhydrous metal sulfates formed.

The specific enthalpies for selected sulfation reactions reported in kilojoules per kilogram (kJ/kg) of metal oxide M_xO_y and ordered in alphabetical order along with the minimum and energetic decomposition temperatures of the resulting anhydrous metal sulfates are reported in Table 31.

[68] HODGMAN, C.D. (ed.)(1962-1963) *Handbook of Chemistry and Physics, 44th Edition*. Chemical Rubber Publishing Co., Cleveland, OH, page 1995.

[69] DUVAL, C. (1963) *Inorganic Thermogravimetric Analysis, Second Edition*. Elsevier Publishing Company, Amsterdam, The Netherlands.

Table 31 – Specific sulfation enthalpy at 180°C and decomposition temperatures

Sulfation chemical reaction schemes	Specific enthalpy of sulfation (kJ/kg)	Decomposition temperature of anhydrous sulfate	
		Minimum	Energetic
$Al_2O_3 + 3H_2SO_4 = Al_2(SO_4)_3 + 3H_2O(g)$	-663	590°C	639°C
$BeO(s) + H_2SO_4 = BeSO_4(s) + H_2O(g)$	-2,988	1000°C	n.a.
$CaF_2(s) + H_2SO_4 = CaSO_4 + 2HF(g)$	+42	1200°C	1400°C
$Ca(OH)_2(s) + H_2SO_4 = CaSO_4 + 2H_2O(g)$	-2,289	1200°C	1400°C
$CaO + H_2SO_4 = CaSO_4 + H_2O(g)$	-4,839	1200°C	1400°C
$Cr_2O_3 + 3H_2SO_4 = Cr_2(SO_4)_3 + 3H_2O(g)$	-714	640°C	n.a.
$Fe_2O_3 + 3H_2SO_4 = Fe_2(SO_4)_3 + 3H_2O(g)$	-379	492°C	560°C
$Fe_3O_4 + 4H_2SO_4 = Fe_2(SO_4)_3 + FeSO_4 + 4H_2O(g)$	-572	See those for FeO and Fe_2O_3	
$FeO + H_2SO_4 = FeSO_4 + H_2O(g)$	-1,756	167°C	480°C
$FeTiO_3(s) + 2H_2SO_4 = FeSO_4(s) + TiOSO_4(s) + 2H_2O(g)$	-4,273	See those for FeO and TiO_2	
$K_2O + H_2SO_4 = K_2SO_4 + H_2O(g)$	-5,758	n.a.	n.a.
$La_2O_3 + 3H_2SO_4 = La_2(SO_4)_3 + 3H_2O(g)$	-3,040	n.a.	n.a.
$Li_2O + H_2SO_4 = Li_2SO_4 + H_2O(g)$	-6,589	1300°C	
$MgO + H_2SO_4 = MgSO_4 + H_2O(g)$	-3,807	998°C	1085°C
$MnO + H_2SO_4 = MnSO_4 + H_2O(g)$	-2,099	690°C	790°C-940°C
$Na_2O + H_2SO_4 = Na_2SO_4 + H_2O(g)$	-7,129	870°C	1100°C
$NiO + H_2SO_4 = NiSO_4 + H_2O(g)$	-1,575	810°C	n.a.
$ThO_2 + 2H_2SO_4 = Th(SO_4)_2 + 2H_2O(g)$	-781	700°C	730°C
$Ti_2O_3 + 3H_2SO_4 = Ti_2(SO_4)_3 + 3H_2O(g)$	-1,261	500°C	600°C
$TiO_2 + H_2SO_4 = TiOSO_4 + H_2O(g)$	-2,029	430°C	800°C
$V_2O_3 + 3H_2SO_4 = V_2(SO_4)_3 + 3H_2O(g)$	-2,480	n.a.	n.a.
$V_2O_4(s) + 2H_2SO_4 = 2VOSO_4(s) + 2H_2O(g)$	-627	520°C	530°C
$V_2O_5 + H_2SO_4 = (VO_2)_2SO_4 + 3H_2O(g)$	-861	n.a.	n.a.
$ZnO + H_2SO_4 = ZnSO_4 + H_2O(g)$	-1,257	735°C	860°C

For instance, during sulfuric acid digestion, it is relatively safe to operate with a maximum targeted baking temperature of 200°C in order to avoid the oxidation of the newly formed metals sulfates with release of noxious SO_2 fumes with traces of SO_3. Actually even if the peak temperature reaches 250°C the duration of that event is too short to observe decomposition.

8.4 Specific Heat Capacity of Materials

The specific heat capacity of the material to be sulfated is required to determine the maximum adiabatic temperature rise during the sulfation reaction.

It can be measured experimentally by calorimetry but if such practical data is not available in the literature, it is always possible to calculate with a reasonable uncertainty from the chemical analysis of the material to be sulfated when the latter is expressed as metal oxides.

The specific heat capacity corresponds then to the weighted averaged of the specific heat capacity for each individual oxide based on the following simple equation:

$$c_p(\text{mixture}) = \Sigma_i w_i c_{pi}$$

With $c_{p\,i}$ specific heat capacity of each oxide in $J.kg^{-1}.K^{-1}$

w_i mass fraction of each individual oxide.

For precise calculations, we can utilise the polynomial form of the molar heat capacities with related empirical coefficients listed in Table 27. However, for quick calculations, it is still possible to use the molar and specific heat capacities listed in Table 32.

Table 32 – Molar and specific heat capacities of metal oxides (fluoride) at 25°C

Metal oxide	J/mol/K	M_R	J/kg/K
Fe_2O_3	103.9	159.7	651
Cr_2O_3	118.7	152.0	781
Al_2O_3	79.0	94.0	841
La_2O_3	109.2	325.8	326
V_2O_5	127.7	181.9	702
P_2O_5	204.8	141.9	1443
SiO_2	44.4	60.1	739
TiO_2	55.0	79.9	689
ThO_2	61.8	264.0	234
BeO	25.6	25.0	1024
CaF_2	57.6	78.1	738
CaO	42.0	56.1	749
MgO	37.2	40.3	923
MnO	45.4	70.9	640
FeO	50.0	71.8	696
NiO		74.7	0
ZnO	40.3	81.4	495
Li_2O	53.5	29.9	1790
Na_2O	69.1	62.0	1115
K_2O	60.3	94.2	640

For instance, we have calculated below based on their chemical composition the specific heat capacity, and the specific enthalpy of sulfation at 453.15K (180°C) for various ores, and concentrates that are processed by sulfuric acid digestion and will be discussed in the following sections.

Table 33 – Specific heat capacities and sulfation enthalpies at 180°C

Mineral, ore, or concentrate	Specific heat capacity [J/(kg.K)]	Specific sulfation enthalpy (kJ/kg)
Bauxite ore (61 wt.% Al_2O_3, 28 wt.% H_2O)	1,762	-489
Fluorspar (97 wt.% CaF_2)	716	+41
Ilmenite (49 wt.% TiO_2)	691	-1,931
BOF-slag (45 wt.% CaO and 10 wt.%MgO)	741	-3,050
Monazite (63 wt.% Ln_2O_3)	649	-680
Phosphate rock (32 wt.% P_2O_5)	965	-1,494
Spodumene (6 wt.% Li_2O)	828	-645
Titano-magnetite (11.4 wt.% TiO_2)	673	1,004

From the few examples listed in the above table, we can see clearly that some materials such as BOF-slag, ilmenite, apatite, monazite, and magnetite when sulfated are strongly exothermic with the release of a large amount of heat enough to maintain an autogenous operation. This is explained by the significant concentration of CaO, FeO, or MgO in these solids. On the other hand, the sulfation of spodumene and bauxite will require an external source of heat to be completed. This behavior is of paramount importance for selecting the proper sulfation method and it will be exemplified in Section 10.

8.5   Maximum Adiabatic Temperature Rise during Sulfation

From the value of the specific enthalpy for the sulfation reaction, it is then possible to calculate the total amount of heat released per unit mass of the mixture, that is, per unit mass of material to be sulfated and sulfuric acid at a given strength.

Afterwards, knowing the specific enthalpy of sulfation together with the initial (A-to-S) mass ratio of concentrated sulfuric acid, and the specific heat capacities of concentrated sulfuric acid, water and of the material respectively it is then possible to calculate the adiabatic temperature rise of the mixture using the following simple equation:

$$\Delta T_{adiabat} = \Delta h_{sulfation} / \{c_{psolid} + [\text{A-to-S}][w_A c_{pacid} + (1 - w_A)c_{pwater}]/w_A\}$$

with

$\Delta T_{\text{adiabat}}$	adiabatic temperature rise, in K,
$\Delta h_{\text{sulfation}}$	specific enthalpy of sulfation, in J/kg,
c_{psolid}	specific heat capacity of the solid, in J/(kg.K),
c_{pacid}	specific heat capacity of sulfuric acid, in J/(kg.K),
c_{pwater}	specific heat capacity of water, in in J/(kg.K),
w_{A}	mass percentage of sulfuric acid in the solution,
[A-to-S]	100% sulfuric acid-to-solid mass ratio.

From the above equation, it is then possible to establish several tables listing the peak temperature under adiabatic conditions for various starting temperatures and acid numbers (Table 34).

Table 34 – Adiabatic temperature rise and practical sulfuric acid numbers

Mineral, ore, or concentrate	Mass percentages of sulfuric acid	Practical sulfuric acid number [H_2SO_4-to-S]	Adiabatic temperature rise (ΔT)
Bauxite ore (61 wt.% Al_2O_3 and 28 wt.% moisture)	78 wt.% (initial) 78 wt.% (attack)	1,867	-260 K (endothermic)
Fluorspar (97 wt.% CaF_2)	93 wt.% (initial) 78 wt.% (attack)	1,219	-97 K (endothermic)
Ilmenite (49 wt.% TiO_2)	85 wt.% (initial) 80 wt.% (attack)	1,541	+273 K
BOF-slag (55 wt.% CaO + 10 wt.% MgO)	93 wt.% (initial) 93 wt.% (attack)	4,588	+297 K
Monazite (63 wt.% Ln_2O_3)	93 wt.% (initial) 90 wt.% (attack)	1,906	+212 K
Perovskite (50 wt.% TiO_2)	80 wt.% (initial) 80 wt.% (attack)	16,127	-215 K
Phosphate rock (32 wt.% P_2O_5)	73 wt.% (initial) 64 wt.% (attack)	833	+141 K
Spodumene (6 wt.% Li_2O)	93 wt.% (initial) 93 wt.% (attack)	317	+331 K
Titano-magnetite (11.4 wt.% TiO_2)	93 wt.% (initial) 90 wt.% (attack)	1,584	+216 K

As expected form the large values for the specific heat capacities of sulfuric acid (1,399 J/kg/K) and water (4,184 J/kg/K), higher is the acid number lower is the adiabatic peak temperature.

8.6 Maximum and Safe Operating Temperature

Three major factors influence the maximum and safe operating temperature during the exothermic sulfation reaction: (1) the corrosion rate of the construction material; (2) the boiling point of the sulfuric acid for a given strength; (3) the vapor pressure of SO_3 above the mixture.

The corrosion rate of the materials in direct contact with the concentrated sulfuric acid is of paramount importance for the dimensional and structural integrity of the reactors, digesters, and tanks.

Because during sulfation roasting and sulfuric acid digestion the concentration of sulfuric acid utilized ranges usually from 80 wt.% up to 98 wt.%, the maximum operating temperature can be close to the boiling point of the concentrated sulfuric acid, that is, ca. 320°C with 98 wt.% H_2SO_4. The boiling point as a function of the mass percentage of sulfuric acid are depicted **Figure 9**.

However, another important consideration must be taken into account when selecting the maximum operating temperature; it is the vapor pressure of water vapor and sulfur trioxide above the concentrated acid. Actually, it is important to avoid losing sulfuric acid value and also to prevent the evolution of noxious fumes that will pose occupational health and safety hazards for the workers.

8.7   Hydration Reaction and Specific Enthalpy of Dilution

It is well known even for the layman, that when mixing concentrated sulfuric acid with water a large amount of heat is released due to the strongly exothermic reaction of hydration which always results in a significant temperature elevation of the mixture.

We can summarize the hydration of sulfuric acid of a given strength with water as follows:

$$(aH_2SO_4 + bH_2O) + cH_2O = [aH_2SO_4 + (b+c)H_2O] + Q_H$$

More generally, we can also express the dilution of sulfuric acid of a given strength w_A with sulfuric acid having a different strength w_B, in order to obtain a final sulfuric acid of strength w_A, according to the following reaction scheme:

$$(aH_2SO_4 + bH_2O) + (cH_2SO_4 + dH_2O) = [(a+b)H_2SO_4 + (c+d)H_2O] + Q_H$$

With the following relationships between the stoichiometric coefficients and the mass percentages:

$$w_A = aH_2SO_4/(aH_2SO_4 + bH_2O)$$

$$w_B = cH_2SO_4/(cH_2SO_4 + dH_2O)$$

$$w_C = (a+b)H_2SO_4/[(a+b)H_2SO_4 + (c+d)H_2O]$$

Numerous data points for the specific enthalpy of sulfuric acid-water solutions that have been obtained by calorimetric experiments have been published.

For convenience, we have reported in Table 35, the specific enthalpy of solutions of sulfuric acid and water including water vapor gathered from several scientific and technical sources. These are reported in both kJ/kg and Btu/lb vs. the mass percentage of sulfuric acid and the temperature.

Table 35 – Specific enthalpy of hydration vs. mass percentage and temperature

		0	5	10	15	20	25	30	35	40	45	50	55	60	65	70	75	80	85	90	95	100
$w_{H_2SO_4}$	wt %	0	5	10	15	20	25	30	35	40	45	50	55	60	65	70	75	80	85	90	95	100
w_{H_2O}	wt %	100	95	90	85	80	75	70	65	60	55	50	45	40	35	30	25	20	15	10	5	0
m_{SOLN}	kg	1.000	1.000	1.000	1.000	1.000	1.000	1.000	1.000	1.000	1.000	1.000	1.000	1.000	1.000	1.000	1.000	1.000	1.000	1.000	1.000	1.000
m_{H_2O}	kg	1.000	0.950	0.900	0.850	0.800	0.750	0.700	0.650	0.600	0.550	0.500	0.450	0.400	0.350	0.300	0.250	0.200	0.150	0.100	0.050	0.000
$m_{H_2SO_4}$	kg	0.000	0.050	0.100	0.150	0.200	0.250	0.300	0.350	0.400	0.450	0.500	0.550	0.600	0.650	0.700	0.750	0.800	0.850	0.900	0.950	1.000
n_{H_2O}	mol	55.51	52.73	49.96	47.18	44.41	41.63	38.86	36.08	33.31	30.53	27.75	24.98	22.20	19.43	16.65	13.88	11.10	8.33	5.55	2.78	0.00
$n_{H_2SO_4}$	mol	0.00	0.51	1.02	1.53	2.04	2.55	3.06	3.57	4.08	4.59	5.10	5.61	6.12	6.63	7.14	7.65	8.16	8.67	9.18	9.69	10.20
n_w/n_a			103.4	49.0	30.9	21.8	16.3	12.7	10.1	8.2	6.7	5.4	4.5	3.6	2.9	2.3	1.8	1.4	1.0	0.6	0.3	0.0
Specific heat capacity (Btu/lb/°F)		0.999	0.966	0.933	0.900	0.866	0.833	0.800	0.767	0.734	0.700	0.667	0.634	0.601	0.567	0.534	0.501	0.468	0.435	0.401	0.368	0.335
BOILING POINT		212°F	214°F	216°F	218°F	220°F	224°F	226°F	230°F	240°F	248°F	260°F	270°F	284°F	304°F	330°F	363°F	399°F	442°F	500°F	550°F	620°F
SPECIFIC ENTHALPY OF SOLUTION																						
ACID	Btu/lb H₂SO₄		-340.4	-308.9	-306.27	-301	-297.8	-287.8	-281.7	-271.3	-262.3	-249.5	-238.3	-223.8	-208.4	-189	-167.4	-144.1	-118.5	-98	-53.77	0
ACID	kJ/kg H₂SO₄		-792	-719	-712	-700	-693	-669	-656	-631	-610	-580	-554	-520	-485	-440	-389	-335	-271	-205	-125	0
SOLUTION 32°F	Btu/lb solution	0.00	-17.82	-30.89	-48.94	-66.28	-74.46	-86.34	-96.51	-108.51	-118.02	-124.79	-131.99	-134.26	-135.46	-132.28	-126.84	-116.26	-98.08	-79.20	-61.08	0.00
SOLUTION 32°F	kJ/kg solution	0	-40	-72	-107	-140	-173	-201	-229	-252	-275	-290	-305	-312	-315	-308	-292	-268	-230	-184	-119	0
SOLUTION 54°F	Btu/lb solution	18	0	-14	-30	-48	-59	-72	-85	-95	-105	-113	-120	-123	-125	-123	-117	-107	-91	-72	-44	5
SOLUTION 54°F	kJ/kg solution	42	1	-33	-69	-104	-138	-167	-197	-222	-245	-262	-278	-287	-291	-286	-271	-248	-212	-167	-103	14
SOLUTION 75°F	Btu/lb solution	43	25	9	-7	-23	-39	-52	-66	-77	-88	-96	-104	-108	-111	-109	-104	-95	-80	-62	-35	14
SOLUTION 75°F	kJ/kg solution	100	57	21	-17	-53	-80	-121	-153	-179	-204	-223	-242	-252	-258	-254	-242	-221	-187	-144	-82	33
SOLUTION 100°F	Btu/lb solution	68	49	33	15	-1	-18	-32	-46	-59	-70	-79	-88	-93	-97	-96	-91	-83	-69	-52	-26	23
SOLUTION 100°F	kJ/kg solution	158	113	76	35	-3	-41	-74	-108	-136	-164	-185	-205	-217	-225	-223	-213	-194	-162	-121	-61	53
SOLUTION 125°F	Btu/lb solution	93	73	56	38	20	3	-12	-27	-40	-53	-63	-72	-78	-83	-83	-79	-72	-59	-42	-17	31
SOLUTION 125°F	kJ/kg solution	216	169	130	88	47	7	-28	-64	-94	-123	-146	-168	-182	-192	-192	-184	-167	-136	-97	-39	72
SOLUTION 150°F	Btu/lb solution	118	97	79	60	42	24	8	-8	-22	-35	-46	-56	-63	-68	-69	-66	-60	-48	-32	-8	40
SOLUTION 150°F	kJ/kg solution	274	226	184	140	98	55	19	-19	-51	-82	-107	-131	-147	-159	-161	-155	-140	-111	-74	-18	92
SOLUTION 175°F	Btu/lb solution	143	121	103	83	64	45	28	11	-4	-18	-29	-40	-48	-54	-56	-54	-48	-37	-22	2	48
SOLUTION 175°F	kJ/kg solution	332	282	238	192	148	104	66	26	-8	-42	-68	-94	-113	-126	-130	-125	-112	-86	-51	4	111
SOLUTION 200°F	Btu/lb solution	168	145	125	105	85	66	48	30	15	0	-13	-25	-33	-40	-43	-41	-37	-26	-12	11	56
SOLUTION 200°F	kJ/kg solution	391	338	293	245	199	152	112	70	34	-1	-29	-57	-76	-93	-99	-96	-85	-60	-27	25	131
SOLUTION 212°F	Btu/lb solution	180	157	137	116	96	76	58	39	24	8	-5	-17	-26	-33	-36	-35	-31	-21	-7	15	60
SOLUTION 212°F	kJ/kg solution	418	365	319	270	223	176	134	92	55	19	-11	-40	-61	-77	-84	-82	-72	-48	-16	35	140
SOLUTION 225°F	Btu/lb solution	1163	1091	1022	952	883	814	68	49	33	17	4	-9	-18	-26	-29	-29	-25	-15	-2	20	65
SOLUTION 225°F	kJ/kg solution	2706	2538	2378	2215	2054	1893	158	115	77	40	9	-20	-43	-60	-68	-67	-58	-35	-4	46	150
SOLUTION 250°F	Btu/lb solution	1188	1115	1045	975	905	835	767	698	633	568	21	7	-3	-12	-16	-16	-13	-4	8	28	73
SOLUTION 250°F	kJ/kg solution	2763	2594	2432	2267	2104	1941	1784	1626	1473	1322	48	17	-8	-27	-37	-38	-31	-10	19	68	170
SOLUTION 275°F	Btu/lb solution	1213	1139	1069	997	926	856	787	718	652	586	522	459	12	2	-2	-4	-2	7	18	38	81
SOLUTION 275°F	kJ/kg solution	2821	2650	2496	2319	2155	1990	1831	1671	1516	1362	1215	1069	27	6	-6	-4	-2	15	43	89	189
SOLUTION 300°F	Btu/lb solution	1238	1163	1092	1020	948	878	807	737	670	603	539	475	415	356	11	9	10	17	28	48	90
SOLUTION 300°F	kJ/kg solution	2879	2706	2540	2372	2205	2038	1877	1715	1569	1403	1254	1106	965	828	25	20	24	41	66	111	209
SOLUTION 325°F	Btu/lb solution	1263	1188	1115	1042	970	897	827	757	688	621	556	491	430	370	315	21	22	28	38	57	98
SOLUTION 325°F	kJ/kg solution	2937	2762	2595	2424	2255	2087	1924	1760	1601	1444	1293	1142	1000	861	733	49	51	66	89	132	228
SOLUTION 350°F	Btu/lb solution	1288	1212	1139	1065	991	918	847	776	707	638	572	507	445	384	329	276	33	39	48	66	106
SOLUTION 350°F	kJ/kg solution	2995	2818	2649	2476	2306	2135	1970	1804	1644	1484	1331	1179	1034	894	764	643	78	91	113	153	248
SOLUTION 375°F	Btu/lb solution	1313	1236	1162	1087	1013	938	867	795	725	656	589	523	460	399	342	289	45	50	58	75	115
SOLUTION 375°F	kJ/kg solution	3054	2875	2703	2529	2356	2184	2017	1849	1687	1525	1370	1216	1069	927	795	672	105	116	136	175	267
SOLUTION 400°F	Btu/lb solution	1338	1260	1185	1110	1035	960	887	814	743	673	606	539	475	413	355	301	251	61	68	84	123
SOLUTION 400°F	kJ/kg solution	3112	2931	2757	2581	2407	2232	2063	1894	1729	1569	1409	1253	1104	960	826	701	584	142	159	196	287
SOLUTION 425°F	Btu/lb solution	1363	1284	1209	1132	1056	980	907	833	762	691	622	555	490	427	369	314	263	72	79	94	132
SOLUTION 425°F	kJ/kg solution	3170	2987	2812	2633	2457	2281	2110	1938	1772	1607	1448	1290	1139	993	858	730	611	167	183	218	306
SOLUTION 450°F	Btu/lb solution	1388	1308	1232	1155	1078	1001	927	852	780	708	639	570	505	441	382	326	274	228	83	103	140
SOLUTION 450°F	kJ/kg solution	3228	3043	2866	2686	2507	2329	2156	1983	1815	1647	1487	1327	1174	1028	889	759	638	531	206	239	326
SOLUTION 475°F	Btu/lb solution	1413	1332	1255	1177	1100	1022	947	872	798	726	656	586	520	455	395	339	286	239	99	112	148
SOLUTION 475°F	kJ/kg solution	3286	3099	2920	2738	2558	2378	2203	2027	1857	1688	1525	1364	1209	1059	920	788	665	556	229	260	345
SOLUTION 500°F	Btu/lb solution	1438	1367	1279	1200	1121	1042	967	891	817	743	672	602	535	470	409	351	298	250	206	121	157
SOLUTION 500°F	kJ/kg solution	3344	3155	2974	2790	2608	2428	2249	2072	1900	1729	1564	1400	1244	1092	951	817	692	581	478	282	365
SOLUTION 525°F	Btu/lb solution	1463	1381	1302	1222	1143	1064	987	910	835	761	689	618	550	484	422	364	309	261	216	130	165
SOLUTION 525°F	kJ/kg solution	3402	3212	3029	2843	2659	2474	2296	2116	1943	1769	1603	1437	1279	1125	982	847	720	606	502	303	384
SOLUTION 550°F	Btu/lb solution	1488	1405	1325	1245	1165	1085	1007	929	853	778	708	634	565	498	435	376	321	272	226	188	173
SOLUTION 550°F	kJ/kg solution	3460	3268	3083	2895	2709	2523	2342	2161	1985	1810	1642	1474	1314	1159	1013	876	747	632	525	437	403

Note that, the specific enthalpy for pure liquid water is taken arbitrarily equal to 0 kJ/kg at 0°C (0 Btu/lb at 32°F). This convention was adopted, in order to comply with the same datum used in the international steam tables [70]. Thus specific enthalpies from the tabulated data can be used seamlessly with those from the steam tables when water vapor (steam) is involved in the calculations.

[70] KRETZSCHMAR, H.-J.; and WAGNER, W. (2019) *International Steam Tables, Properties of Water and Steam based on the Industrial Formulation IAPWS-IF97, Tables, Algorithms, Diagrams, and Online Calculation, Third Edition.* Springer-Verlag, Berlin, Germany.

It is worth to mention that the maximum heat released upon mixing sulfuric acid and pure water is located at the mass percentage of 66 wt.% H_2SO_4.

These tabulated data allow the calculation of the following thermodynamic quantities: (1) the total amount of heat released upon dilution per unit mass of solution, (2) the adiabatic temperature rise during dilution, and finally (3) the amount of heat that must be removed to bring back the temperature of the solution down to a selected temperature. These calculations are detailed hereafter.

8.7.1 Specific Amount of Heat Released by Dilution

If a concentrated sulfuric acid of mass m_A, and strength a is diluted with a mass m_B of diluted acid of strength b, in order to obtain a mass m_C of strength c, according to the mixing rules (Section 18.3), the three relations between these masses are as follows:

$$m_A = m_C \, (c - b)/(a - b)$$

$$m_B = m_C \, (a - c)/(a - b)$$

$$m_B = m_A \, (a - c)/(c - b)$$

On the other hand, knowing the three specific enthalpies of each substance at a given temperature, the conservation of energy imposes that after dilution the total amount heat of dilution released Q_H in kJ (Btu) is given by the following mass and energy balance equation:

$$Q_H = (m_C \, h_C - m_A \, h_A + m_B \, h_B)$$

Therefore the specific heat of hydration released per unit mass of solution is given by:

$$q_H = Q_H/m_C = (m_C \, h_C - m_A \, h_A + m_B \, h_B)/m_C$$

or by replacing the masses by the mixing equations

$$q_H = h_C - [(c - b)/(a - b)] \, h_A + [(a - c)/(a - b)] \, h_B$$

Example: if we need to prepare 1,000 kg (2,205 lb) of an aqueous solution of 66 wt.% sulfuric acid, with a specific enthalpy of −257 kJ/kg (-110 Btu/lb), by diluting 96 wt.% concentrated, with a specific enthalpy of −58.6 kJ/kg (-25 Btu/lb), with deionized, +100 kJ/kg (+43 Btu/lb) both reactants being at 75°F (ca. 23.9°C) based on the mixing rules, we need to mix 687.5 kg (1,516 lb) of 66 wt.% H_2SO_4 with 312.5 kg (689 lb.) of deionized water. Therefore, the total specific heat released by dilution per unit mass of solution is: -257 x 1000 - (-58.6 x 687.5 + 100 x 312.5) = -248 kJ/kg (-107 Btu/lb).

8.7.2   Adiabatic Temperature Rise upon Dilution

The maximum adiabatic temperature rise $\Delta T_{adiabat}$ expected from the dilution of concentrated sulfuric acid can be approximated based on the specific heat capacities of pure sulfuric acid c_{acid} (1.399 kJ.kg^{-1}.K^{-1}) and that of water c_{water} (4.184 kJ.kg^{-1}.K^{-1}), and the final strength of the sulfuric acid expressed as mass percentage w_C as follows:

$$\Delta T_{adiabat} = q_H / [w_C c_{pacid} + (1 - w_C) c_{pwater}]$$

It is important to mention that unless the vessel in which the dilution is performed is perfectly insulated thermally, the actual temperature rise is always lower than the adiabatic temperature rise.

Example: from the previous example, we obtained the specific heat of dilution -248 kJ/kg (-107 Btu/lb) for preparing an aqueous solution of sulfuric acid with 66 wt.%. H_2SO_4. The specific heat capacity of the mixture is: (0.66 x 1.400 + 0.34 x 4.184) = 2.347 kJ.kg^{-1}.K^{-1}. Therefore, the maximum adiabatic temperature rise is ca. 106 K. This mean, that theoretically, inside an insulated vessel, the final temperature of the mixture can reach as high as 130°C (266°F).

8.7.3   Removal of Heat by Cooling

As exemplified in the previous paragraphs, due to the strongly exothermic reaction of hydration, the maximum temperature of the final solution is often very high especially when diluting with water concentrated sulfuric acid with a strength greater than 80 wt.%. H_2SO_4.

Therefore, in order to avoid the instant boiling of the solution with catastrophic failure of plastic tanks and polymer vessels or to prevent thermal runaways, it is mandatory to remove continuously the evolved heat by cooling with immersed coils, or by means of cross-flow or plate and frame heat exchangers.

The total amount of heat per unit mass of solution to be removed corresponds simply to the difference between the specific enthalpy of the solution at the final temperature, $h_C(T_{max})$, minus the specific enthalpy of the solution at a lower temperature $h_C(T_0)$.

$$q = Q / m_C = h_C(T_{max}) - h_C(T_0)$$

Example: if we need to maintain for safety reasons, a maximum temperature of 100°F (ca. 37.8°C), during the preparation by dilution of 66% sulfuric acid with the same reactants and operating conditions described in the previous examples, we will need to remove: 4.65 − (-227.94) = 232 kJ/kg (99.7 Btu/lb).

8.7.4　Graphical Method

In practice, it is often easier and faster to utilise a graphical method with sufficient accuracy for the calculation of the heat released upon dilution of sulfuric acid, the maximum adiabatic temperature rise, and the heat to be removed.

Figure 14 depicts the specific enthalpy of solution-mass percentage diagram for sulfuric acid and water mixtures. The curved plots represent the isotherms.

At this point, it is important to note that from the close examination of the graph, the maximum amount of heat is released when preparing 66 wt.% sulfuric acid.

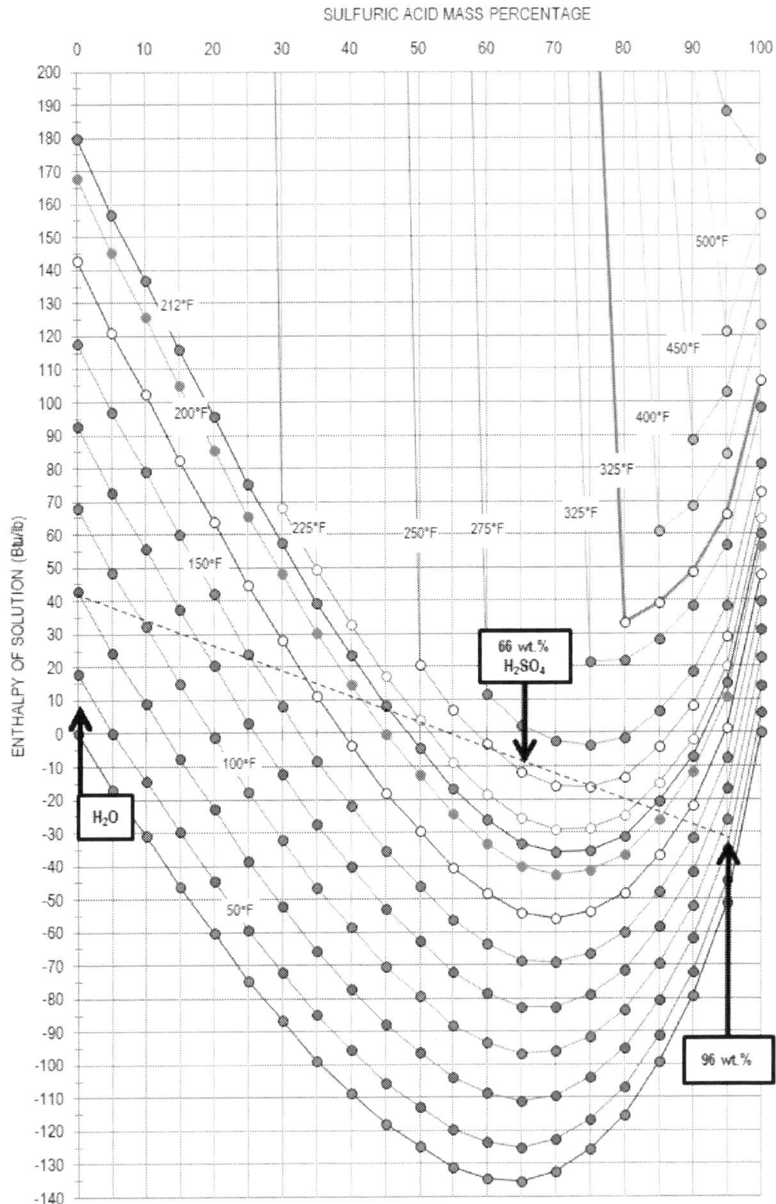

Figure 14 – Specific enthalpy of solution vs. mass percentage of acid

<u>Conversion factors</u>: 1 Btu(therm.)/lb. $\approx$ 2.325 kJ/kg, and $t(°C) = (5/9)[t(°F) - 32]$

For instance, under the same operating conditions used in the previous examples, if we draw a straight line from the coordinates for pure liquid water to those of 96% sulfuric acid both at 75°F (23.9°C), we can extrapolate directly on the graph from the intersecting line and the closest isotherm for an abscissa corresponding to a mass percentage of 66%, the adiabatic temperature obtained graphically is about 266°F (130°C).

Afterwards, the amount of heat to be removed to bring the temperature down to 100°F (23.9°C) is determined by the difference along the vertical line between the specific heat of solution minus the specific heat at 100°F that is -251 kJ/kg (-108 Btu/lb).

9 Industrial Sulfation Processes

9.1 Titanium Dioxide by the Sulfate Process

The sulfate process was introduced commercially in 1916 by the *National Lead Company*[71] now *Kronos* and it is still extensively used today across all the titanium pigment industry worldwide. Originally, the **NL Process** was restricted to the sulfation of beach sand ilmenite with a titanium dioxide content ranging from 52 to 54 wt.% TiO_2. The entire process for the sulfation of ilmenite has been extensively described by Barksdale [72] at that time research chemist at the Titanium division of the *National Lead Company*.

Beach sand ilmenite suitable for the sulfation process originates mainly from Australia, South Africa, the USA, Brazil, and Ukraine with an exception of hard rock ilmenite mined from Norway.

Following the pioneering work of Peirce et al. [73] in the 1940s, that demonstrated the possibility to produce titanium-rich slags by smelting hard rock ilmenite and hemo-ilmenite unsuitable for the sulfate process *New Jersey Zinc* and *Kennecott Utah Copper Co.* (KUCC) started in the 1950s the construction of the metallurgical complex of *Quebec Iron & Titanium* (QIT) located in Sorel-Tracy, Qc, Canada (now *Rio Tinto Fer et Titane Inc.*).

The metallurgical combine became the largest titanium smelter in the world, and started the large scale production of titanium slag from the smelting of hard rock hemo-ilmenite mined from the anorthosite Lac Tio deposit, located in Allard Lake, Qc, Canada [74]. During the smelting and slagging process of hard

[71] MORAN, W.G (1942) *Manufacture of titanium dioxide pigments*. U.S. Patent 2,278,709 (National Lead Company), April 7th, 1942.

[72] BARKSDALE, J. (1966) *Titanium: Its Occurrence, Chemistry and Technology, Second Edition*. The Ronald Press Company, New York, NY, pp. 117-201.

[73] PEIRCE, W.M.; WARING, R.K.; and FETTEROLF, R.K (1947) Titaniferous material for producing titanium dioxide. U.S. Patent 2,476,453 (Quebec Iron and Titanium Corp.), August 17th, 1947

[74] HAMMOND, P.(1952) Allard lake ilmenite deposits. *Economic Geology*, **47**,634-649.

rock ilmenite with anthracite coal inside an electric arc furnace (EAF) [75], the titanium slag with 80 wt.% TiO_2 also called in short **sulfate slag** is co-produced along with high quality pig iron. Thus it rapidly gained acceptance among all the TiO_2 producers using the sulfate process due to its high titanium content and low concentration of iron.

In 2018, the annual production of sulfate slag and beach sand ilmenite suitable for the sulfate process represented about half of all the TiO_2 produced with almost 4 million tonnes per year when expressed as TiO_2 units.

For instance, the average chemical compositions of typical beach sand ilmenite and sulfate titanium slag are reported in Table 36.

Table 36 – Chemical composition of beach sand ilmenite and titanium slag

Oxide(wt.%)	Beach sand ilmenite	Titanium slag (Sorelslag®)
TiO_2 (tot al)	46.95	80.00
Ti_2O_3	nil	18.00
FeO	34.70	9.70
Fe_2O_3	8.35	nil
MgO	3.12	5.30
SiO_2	3.00	2.60
Al_2O_3	1.10	2.40
CaO	1.00	0.50
MnO	0.69	0.30
V_2O_5	0.11	0.60
Cr_2O_3	0.05	0.20

A comprehensive study, experimental trials, and thermodynamic calculations were conducted by Turgeon [76] at the prototype scale in an effort to optimise the process and develop a better understanding of the sulfation of titanium-slags especially the Sorelslag® (Trademark of *Rio Tinto Fer et Titane Inc.*) and that was supported by a thorough design of experiments. The prototype reactor was built to simulate the operation of an industrial digester.

Generally speaking, prior the sulfuric acid digestion, the titanium-rich feedstock either ilmenite or titanium slag is first dried to a moisture content of less

[75] GUEGUIN, M.; and CARDARELLI, F. (2007) Chemistry and mineralogy of titania-rich slags. Part 1 - hemo-ilmenite, sulphate, and upgraded titania slags. *Mineral Processing and Extractive Metallurgy Review (MP&EMR)*, **28**(1)1-58.

[76] TURGEON, J.-F. (1992) *Digestion of titaniferous slags with sulphuric acid*. Master Engineering Thesis, Department of Mining and Materials Engineering, McGill University, Montreal, QC, Canada.

than 0.1 wt.%. The dried raw material is then ground inside a ball mill down to a mean particle size of about 40 µm (minus 325 mesh).

When titanium slag is used instead of ilmenite, the minute amount of free metallic iron originating from the smelting process and embedded as tiny particles along with tramp iron is removed by means of low intensity magnetic separation (LIMS) down to an acceptable upper threshold of 0.5 wt.% Fe(metal).

The latter is mandatory in order to prevent the hazardous evolution of nascent hydrogen gas during the subsequent digestion with sulfuric acid that could reach if not properly addressed the hydrogen lower explosive limit (LEL of 4.0 vol.%) inside the digester headspace.

The two schematic flow diagrams describing the sulfation of ilmenite and the sulfation of titanium slags are depicted in Figure 15 and in Figure 16 respectively.

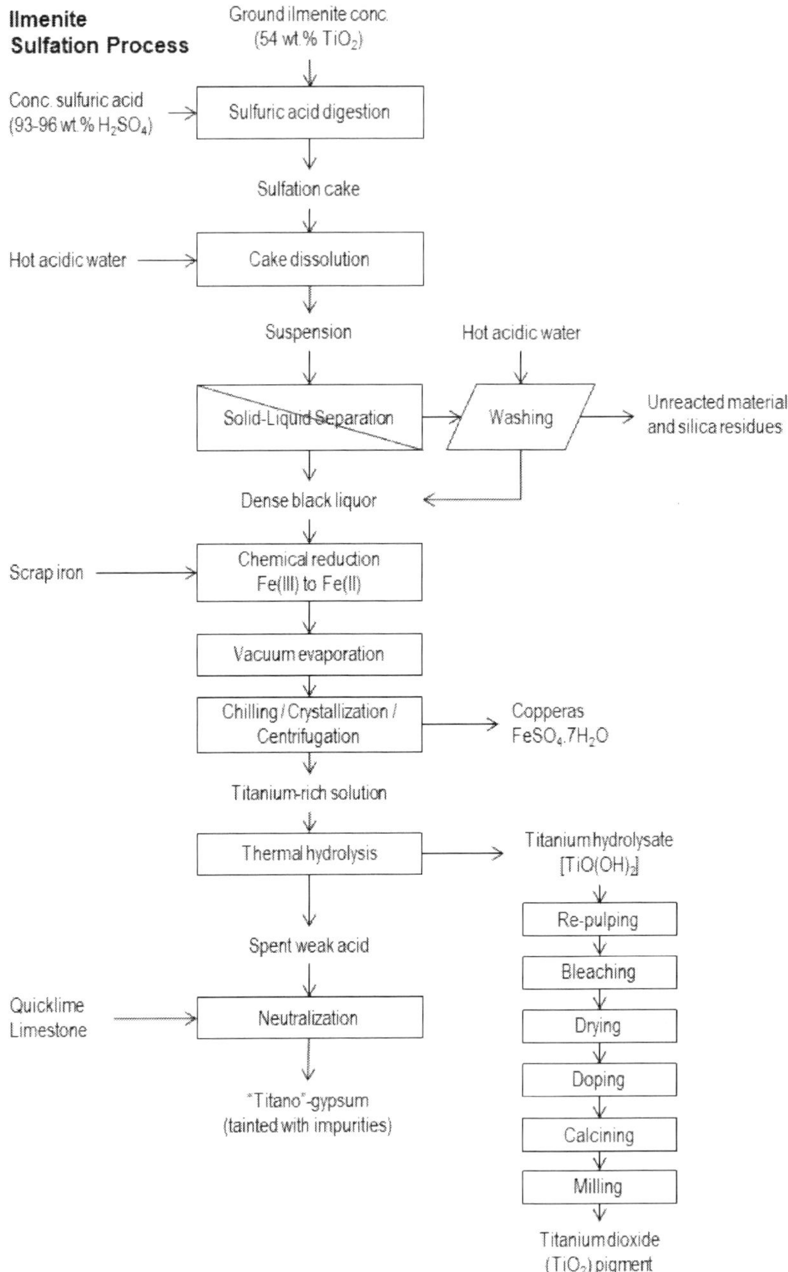

Figure 15 – Flow diagram for the production of titanium dioxide from ilmenite

The batch digestion is then performed inside tall brick lined steel vessel with a conical bottom called *digester* some reaching a brim capacity up to 140 m³. The ground material (e.g., ilmenite, titanium slag, or a blend) is mixed with concentrated sulfuric acid ranging from 85 wt.% up to 98 wt.% H_2SO_4.

The 93 wt.% strength called the "Winter acid" is used essentially in Northern climate regions of the USA, Canada, Russia, and Scandinavia (Section 4.1) avoiding freezing issue during transport and handling.

The targeted sulfuric acid number, that is, the mass ratio of sulfuric acid-to-raw material is chosen such as that the final ratio of free remaining sulfuric acid [H_2SO_4-to-TiO_2] in the subsequent hydrolysis step is about 1,800 kg (100 wt.% H_2SO_4) per tonne of titanium dioxide.

When using highly reactive ilmenites, the strength of sulfuric acid used can be low as around 85 wt.% H_2SO_4. For a beach sand ilmenite with a chemical composition reported in Table 37, the stoichiometric sulfuric acid-to-solid mass ratio [H_2SO_4-to-S] is nearly 1,340 kg (100 wt.% H_2SO_4) per tonne. Thus when using a 15% mass excess the actual sulfuric acid-to-solid mass ratio [H_2SO_4-to-S] is 1,541 kg (100 wt.% H_2SO_4) per tonne of ilmenite, which is equivalent to a practical (actual) sulfuric acid-to-solid mass ratio [A-to-S] of 1,813 (85 wt.% H_2SO_4) per tonne prior water injection. The resulting thick dense slurry exhibits a solid pulp density of 35 mass percent solids.

The suspension is agitated by blowing compressed air from the bottom of the digester while heating is supplied by simply injecting superheated steam into the mixture which is held at a temperature ranging from 70°C to 80°C.

Then water is injected into the slurry to reach a desired strength of sulfuric acid during attack. This results in a steep temperature increase in a matter of few minutes. This first exothermic reaction is only due to the enthalpy of dilution of the concentrated sulfuric acid.

For instance, if water at 20°C (68°F) is injected into 85 wt.% H_2SO_4 at 85°C (185°F) resulting in a final sulfuric acid concentration of 80 wt.% during the attack, the specific enthalpy released per unit mass of final solution is rather significant with -51 kJ/kg (-21.9 Btu/lb).

This translates into an adiabatic temperature rise of 33°C (91°F) and a final temperature of 118°C (276°F), in practice due to unescapable heat losses and sensitive heat absorbed the solids and the acid proof refractory bricks, the temperature is rather around 90°C-95°C.

In the particular case of ilmenites, because of the strong exothermic sulfation reaction that can become explosive if not controlled properly, the heating stage by blowing steam is usually omitted, and only the addition of water is sufficient to trigger the sulfation reaction.

The key figures related to the sulfation of ilmenite are summarized and reported in Table 37 hereafter.

Table 37 – Sulfation of ilmenite: key figures

BEACH SAND ILMENITE (conc. 47 wt.% TiO2)			
Chemical composition		Strength H_2SO_4 (initial)	**85%**
TiO_2	46.95 wt.%	Strength H_2SO_4 (attack)	80%
FeO	34.70 wt.%	[H_2SO_4-to-Solid](Theor.)	**1,340** kg/tonne
Fe_2O_3	8.35 wt.%	Mass excess of acid	**15%**
MgO	3.12 wt.%	[H_2SO_4-to-Solid](Actual)	**1,541** kg/tonne
SiO_2	3.00 wt.%	[H_2SO_4-to-TiO_2](Actual)	**3,283** kg/tonne
Al_2O_3	1.10 wt.%	[Acid-to-Solid](Actual)	**1,813** kg/tonne
CaO	1.00 wt.%	Solids pulp density	36%
MnO	0.69 wt.%	c_p (solid)	692 J/kg/K
V_2O_5	0.11 wt.%	c_p (acid)	1554 J/kg/K
	99.02 wt.%	c_p (pulp)	**1265** J/kg/K
		$\Delta h_{Sulfation}$	**-1787** kJ/kg of solid
Mass of solids (dry)	**1000** kg	$\Delta h_{Hydration}$	-51 kJ/kg of acid
Mass of sulfuric acid	1813 kg	$\Delta h_{Vaporization}$	2256 kJ/kg of steam
Mass of water (injected)	113 kg	Δh_{Total}	**-345** kJ/kg of mixture
Mass of pulp	**2927** kg	Adiabatic temp rise ΔT	**273 K EXOTHERMIC**

This temperature is then sufficient to trigger the exothermic sulfation reaction with a temperature further increasing to about 200°C and that carries on autogenously. It is worth to mention that the sulfation of ilmenites is much more exothermic and violent than that of titanium slags due to the higher reactivity of the former as reported in Table 33.

The mixture is allowed to remain at about 180-200°C for several hours, the so-called acid baking (maturation). Usually due to the low heat losses prevented by the inner refractory lining, and the large thermal inertia of the digester, maintaining the operating temperature at that level for several hours, is not really difficult.

The proper baking temperature is always dictated by: (1) preventing the decomposition of the metal sulfates formed and the evolution of noxious sulfur oxides fumes; (2) avoiding the losses of sulfuric acid by decomposition and intense evaporation, and (3) limiting the number of corrosion resistant materials commercially available for containing such corrosive mixture under harsh conditions.

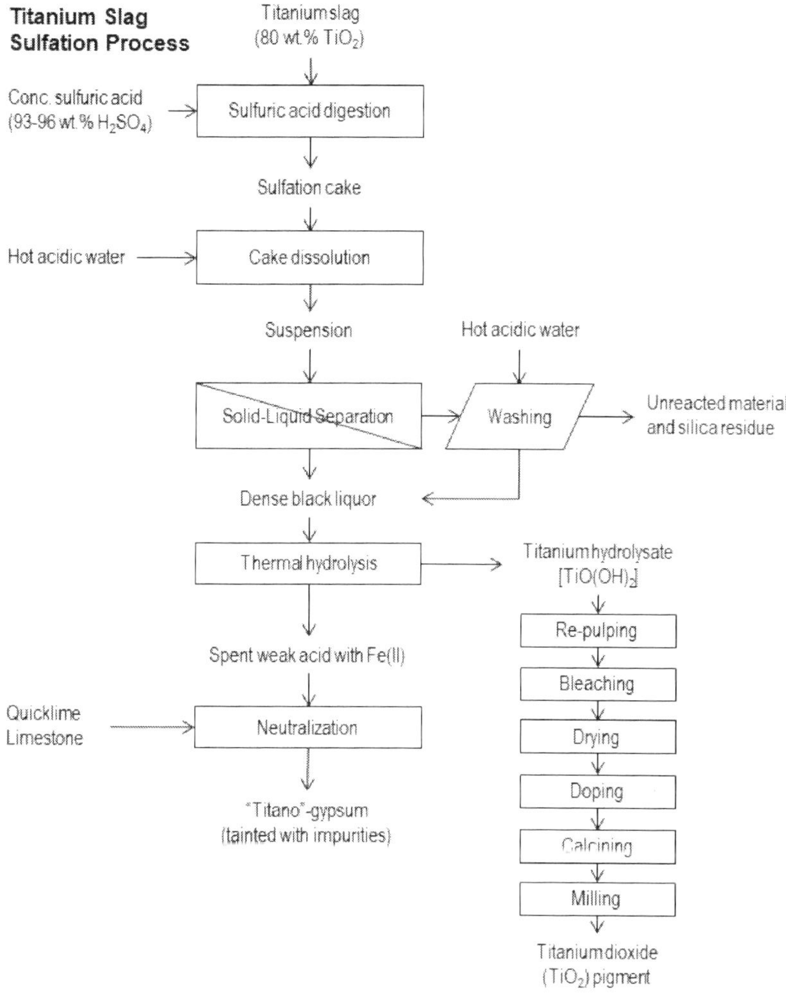

Figure 16 – Flow diagram for the production of titanium dioxide from titanium slag

In the particular case of ilmenites, at least one plant in Brazil originally constructed by the German company *Bayer AG* (now *Tronox*) is using the continuous digestion instead of a batch digestion consisting to feed the ilmenite concentrate, and strong sulfuric acid directly into a screw feeder with a rather short residence time inside the kiln.

After cooling down to 70°C, the solid sulfation cake still at the bottom of the digester is dissolved with warm water for few hours at a temperature below 80°C to avoid premature hydrolysis of Ti(IV). This yields dense and dark liquors.

The solution is then allowed to clarify by settling, and is filtered using a rotary vacuum filter to remove any undissolved residues.

The clear and dense liquor produced from ilmenite always contains some ferric iron that must be reduced using scrap iron metal (Fe) which is added to reduce all the ferric cations according to the redox reaction:

$$Fe(s) + Fe^{3+} = 2Fe^{2+}$$

In the case of titanium slag, the trivalent titanium cations (Ti^{3+}) originating from dissolution of the titanate phases in the slag reduces all the ferric iron (Fe^{3+}) according to the following redox reaction:

$$Ti^{3+} + Fe^{3+} + H_2O = TiO^{2+} + Fe^{2+} + 2H^+$$

It is common practice industrially to leave a slight excess of Ti(III) at a concentration around 3 g/L, to prevent the air oxidation of the ferrous iron in the liquor prior hydrolysis. Actually, the presence of ferric iron is detrimental as it is much more readily adsorbed by the titanium hydrolysate, that is, titanium oxyhydrate ($TiO(OH)_2$) than is ferrous iron, and if ferric iron is present a TiO_2 pigment of good quality cannot be obtained downstream in the process.

The clarified liquor contains about 200-300 g/L of titanyl sulfate ($TiOSO_4$) and about 30-50 g/L of total iron when titanium slag is used.

However, when ilmenite is used as feedstock, up to 120-150 g/L of total iron is present in the solution following the reduction of all the ferric cations by the addition of scrap metallic iron described above. In the latter case, the iron rich solution is cooled down under vacuum to 5°C to allow for the crystallization of iron (II) sulfate heptahydrate ($FeSO_4.7H_2O$), called ***copperas*** or ***melanterite*** when found in nature as a mineral (see Section 11.1).

The actual copperas by-product is in fact more complex as it consists of a waste which is essentially made of wet crystals contaminated with other metal sulfates if the washing of the copperas crystals was not properly done. Thus only clean ferrous sulfate heptahydrate is commercialized. When titanium slags are used, the amount of copperas is small enough, that it ends-up with the weak spent acid.

In all cases, the remaining iron-depleted and titanium-rich dense liquor with a mass density above 1,500 kg/m^3 contains about 170-230 g/L TiO$_2$, 20-30 g/L iron (II) and 20-28 wt.% H$_2$SO$_4$.

The titanium hydrolysate, $TiO(OH)_2$, is precipitated by thermal hydrolysis by simply boiling at 100-110°C for several hours using mostly the *Blumenfeld method* and in some instance by seeding the hot solution by freshly precipitated titanium oxyhydrate or seed of ***anatase*** or ***rutile***.

The crude titanium hydrolysate is then vacuum filtered, re-pulped, and bleached with diluted sulfuric acid and pure aluminum or zinc powders or other proprietary organic reducing agents. This, in in order to remove the entrapped ferrous iron and the last traces of other chromophores impurities essentially Cr(III), and V(IV). Then, it is filtered again, oven dried, doped with pigmentary additives, calcined at 800-1,100°C, and finally milled to produce the desired white TiO_2 pigment.

Following the hydrolysis, the spent liquor or weak acid contains about 20 wt.% H_2SO_4 and about 20-30 g/L iron and sometime vanadium, the liquid effluent is usually neutralized with quicklime and limestone to yield the so-called **titanogypsum**, an impure gypsum of no commercial value tainted with metal hydroxides/carbonates that is disposed-off and landfilled (Section 11.2.2).

We have reported in Table 38, the major TiO_2 pigment producers currently using the sulfate process worldwide with their plant locations, their nameplate capacities, and feedstocks according to several sources and available public information gathered in the period 2015-2020. Regarding the Chinese producers due to the plethora of plants built in mainland China, we have only reported the data for the top ten in 2016 [77] and consolidated the data for the numerous other producers.

Table 38 – Major titanium pigment producers using the sulfate process

Company	Country	Plant location	Nameplate (tonnes TiO_2)	Feedstocks
Tronox (formerly Cristal, Millenium Inorganic Chemicals, KerMcGee, Bayer AG)	Brazil	Salvador, Bahia	60,000	Beach sand ilmenite
	France	Thann	30,000	Beach sand ilmenite
	Germany	Uerdingen	105,000	Blend slag/ilmenite
Venator (Formerly Huntsman, Sachtleben Chemie Metallgesellschaft, Kemira Pigment, Tioxide)	France	Calais (Closed)	100,000	Titanium slag
	Finland	Pori (Closed)	130,000	Norwegian ilmenite
	Germany	Duisburg-Homberg	100,000	Titanium slag
	Italy	Scarlino	80,000	Titanium slag
	Spain	Huelva	73,000	Beach sand ilmenite
	South Africa	Umbogintwini	40,000	Titanium slag
	Malaysia	Teluk Kalung	56,000	Beach sand ilmenite
Kronos (formerly NL Chemicals Inc.)	Canada	Varennes	18,000	Titanium slag
	Germany	Leverkusen	35,000	Norwegian

[77] CCM Data & Business Intelligence (2016) *Top 10 Chinese Titanium Dioxide (TiO2) exporters 2016*.

Company	Country	Plant location	Nameplate (tonnes TiO$_2$)	Feedstocks
				ilmenite
	Germany	Nordenham	60,000	Norwegian ilmenite
	Norway	Fredrikstad	32,000	Norwegian ilmenite
Cinkarna	Slovenia	Cejle	47,000	Titanium slag
Precheza (Precolour AS)	Czech Rep.	Prerov	35,000	Beach sand ilmenite
Zaklady Chemiczne Police SA	Poland	Police	41,000	Blend slag/ilmenite
Agrokhim	Ukraine	Sumy	40,000	Beach sand ilmenite
Kryms Titan (Lakokraska)	Ukraine	Armyansk	80,000	Beach sand ilmenite
ISK (Ishihara Sangyo Kaisha)	Japan	Yokkaichi	54,000	Beach sand ilmenite
Titan Kogyo	Japan	Ube	16,800	Beach sand ilmenite
Sakai Chemical Industry Co. Ltd.	Japan	Onahama, Iwaki	60,000	Beach sand ilmenite
Tayca Corporation	Japan	Saidaiji, Okayama	60,000	Blend slag/ilmenite
Kilburn Chemicals	India	Thoothukkudi, Tami	4,000	Beach sand ilmenite
Kolmak Chemicals	India	Kalyani, Calcutta	6,000	Beach sand ilmenite
Travancore Titanium Ltd. (TTP)	India	Thiruvananthapruram	15,000	Beach sand ilmenite
	South Korea	Incheon	36,000	Beach sand ilmenite
Hankook Titanium	South Korea	Onsam	30,000	Beach sand ilmenite
Sichuan Lomon Group Titanium	China	Mianzhu, Sichuan	173,000	Blend slag/ilmenite
Henan Billions Chemicals	China	Jiaozuo City, Henan	107,000	Blend slag/ilmenite
Gansu Huayuan Titanium	China		61,000	Blend slag/ilmenite
Shandong Dongjia Group	China	Zibo, Shandong	100,000	Blend slag/ilmenite
Shangdong Dawn Titanium	China	Longkou, Shangdong	42,000	Blend slag/ilmenite
Jinan Yuxing Chemical	China	Jinan, Shangdong	31,000	Blend slag/ilmenite
Ningbo Xinfu	China	Ningbo, Zhejiang	20,000	Blend slag/ilmenite
Panzhihua Dongfang Titanium	China	Panzhihua, Sichuan	20,000	Blend slag/ilmenite
Jiangxi Tikon Chemical	China	Fuzhou, Jiangxi	15,000	Blend slag/ilmenite
Guangxi Jinmao Titanium	China	Tengxian, Guangxi	15,000	Blend slag/ilmenite

Company	Country	Plant location	Nameplate (tonnes TiO_2)	Feedstocks
Other Chinese producers	China	Various locations	350,000	Blend slag/ilmenite

9.2   Phosphate and Superphosphate

Along with nitrogen and potassium compounds, phosphorus compounds are key ingredient in the manufacture of fertilizers as essential plant nutrient with 90% of phosphate rock used in the fertilizer industry.

However, the utilization of phosphate bearing materials strongly depends on the solubility and availability of elemental phosphorus to the plants. Among phosphate-bearing raw materials called generically ***phosphate rock***, two type of phosphate rocks can be distinguished based on the geological setting of the deposits:

(1) ***Hard rock apatite*** or ***crystalline phosphate rock*** of igneous origin with large deposits of economic importance usually located in ancient cratons such in the South or Murmansk in the Kola Peninsula (Russia), in Norway, and in Quebec (Canada). The ***apatite-rock*** is made essentially of fluoro-apatite $Ca_5[(PO_4)_3(F,Cl,OH)]$ which is difficult to solubilise that explains its limited use. Commercial deposits are currently exploited in Russia and Norway while Quebec pegmatite deposits that were producing prior 1900 see a regain in interest in the last years with projects such as *Arianne Phosphate* in Lac à Paul and *Mine Arnaud* in Sept-Îles.

(2) ***Phosphate rock*** of sedimentary origin called ***phosphorite*** is made of cryptocrystalline calcium phosphate called with the generic term ***collophane*** [78] with largest deposits located in China, Morocco, the United States, and Russia.

According to the *U.S. Geological Survey's* (USGS) [79] most recent data, the top ten phosphate rock-producing countries for 2020 are reported in Table 39.

---

[78] BATES, R.L. (1960) *Geology of Industrial rocks and Minerals*. Harper & Brothers, Publishers, New York, NY, pp. 178-198.

[79] JASINSKI, S.M. (2021) *Phosphate rock*. U.S. Geological Survey, Mineral Commodity Summaries, January 2021, Washington, D.C.

Table 39 – Top ten phosphate producing countries

Country	Annual production (million tonnes)	Deposits and mines locations	Major companies
China	90	Laohudong	Chanhen
Morocco and Western Sahara	37	Benguerir and Youssoufia Jorf Lasfar Phosboucraa Khouribga	Office Cherifien de Phosphate (OCP)
United States	24	Florida Idaho North Carolina Utah	The Mosaic Company Monsanto Nutrient Itafos Conda LLC J.R. Simplot
Russia	13	Oleniy Ruchey Selemdzha Selgdarsky Tomtor	PhosAgro UralChem Uralkali Acron Eurochem
Jordan	9.0	Al Abiad Al Hassa Al Ruselfa Eshidiya	Jordan Phosphate Mines Company (JPMC)
Saudi-Arabia	6.5	Al Jalamid Ras Al Khair	JV between Ma'aden, SABIC and The Mosaic Company
Brazil	5.5	Agnico dos Dias Araxa Catalao Cajati Irece Tapira Patos de minas	B&A Fostato Mineracao CMOC Galvani Itafos e GB Minerais Vale Fertilizantes
Egypt	5.0	Abu Tartour	El Wady for Phosphate Industries and Fertilizers (WAPHCO)
Vietnam	4.7	Bao Thang province	Duc Giang
Tunisia	4.0	Kef Shfaier - Metlaoui Métlaoui Kef Eddour Erg-Lasfar –Redeyef Mrata – Moularès Jellabia and Mzinda	Compagnie des Phosphate de Gafsa (CPG) Groupe Chimique Tunisien (GCT)

The phosphate rock is usually characterized in the industry either by its content of tricalcium phosphate $Ca_3(PO_4)_2$, CAS No. [7758-87-4], called ***bone phosphate of lime*** with the trade acronym BPL or by its content of phosphorus pentoxide (P_2O_5), CAS No. [1314-56-3]. The relations between the two being simply:

mass percent (wt.% P_2O_5) = 0.45762 [mass percent (wt.% BPL)]

mass percent (wt.% BPL) = 2.1852 mass percent (wt.% P_2O_5)

The easiest way to render the phosphorus of these products available is to perform the so-called ***acidulation*** of the phosphate rock concentrates. The most common method introduced originally by Justus von Liebig uses sulfuric acid but some operations in Europe use nitric acid or a mixture of sulfuric and nitric acids instead [80]. Only the former will be addressed here.

The sulfation of phosphate rock to produce the so-called superphosphate was invented by the English farmer John Lawes that consists to perform the sulfation of the ground phosphorite with concentrated sulfuric acid proceeds according to the following reaction scheme yielding monocalcium phosphate, with the chemical formula $CaH_4(PO_4)_2$, CAS No. [10031-30-8], and gypsum:

$$Ca_3(PO_4)_2 + 2H_2SO_4 + 6H_2O = CaH_4(PO_4)_2 + 2(CaSO_4.2H_2O) + 2H_2O$$

Of course because the phosphate rock always contains fluoroapatite, the latter always reacts with the strong sulfuric acid evolving hydrogen fluoride (HF) gas during the acidulation process.

$$CaF_2(s) + H_2SO_4(l) + 2H_2O(l) = CaSO_4.2H_2O(s) + 2HF(g)$$

However, HF itself reacts with silicates from the gangue to yield the volatile silicon tetrafluoride, SiF_4, CAS No. [7783-61-1], according to the following chemical reaction:

$$4HF(g) + SiO_2(s) = SiF_4(g) + 2H_2O(l)$$

Finally, the silicon tetrafluoride, is hydrolysed into fluorosilicic acid (see also Section 9.7) leaving behind a silica residue:

$$3SiF_4(g) + 2H_2O(l) = 2H_2SiF_6(g) + 3SiO_2(s)$$

---

[80] YOUNG, R.D.; and DAVIS, C.H. (1980) Phosphate Fertilizers and Process Technology. In KHASAWNEH, F.E.; SAMPLE, E.C.; and KAMPRATH, E.J. (eds.) (1980) *The Role of Phosphorus in Agriculture.* Published by American Society of Agronomy, Inc. Crop Science Society of America, Inc. Soil Science Society of America Inc., John Wiley & Sons, New York, NY, pp. 195-226.

Therefore, the overall acidulation reaction can be summarized using the theoretical stoichiometric formula for fluoroapatite as:

$$Ca_{10}(PO_4)_6F_2 + 7H_2SO_4(l) + 3H_2O(l) = 3CaH_4(PO_4)_2.H_2O(s) + 2HF(g) + 7CaSO_4$$

The above intimate mixture of 45 wt.% monocalcium phosphate monohydrate with the chemical formula $CaH_4(PO_4)_2.H_2O$, CAS No. [10031-30-8], and 55 wt.% phosphogypsum is called ordinary or single **superphosphate** in the trade with a common NPK grade or label of 0-22-0 according to the international labelling nomenclature adopted by the fertilizer industry (i.e., 0 wt.% nitrogen (N), 22 wt.% phosphorus (P) expressed as phosphorus pentoxide, and 0 wt.% potassium (K) expressed as potassium oxide).

Because, the monocalcium phosphate is enough soluble into water it is able to supply efficiently phosphorus to crops. However, because the single superphosphate contains up to 55 wt.% inert gypsum this reduces significantly the available concentration of monobasic calcium phosphate.

Based on the above sulfation reaction 1,000 kg of fluoroapatite will yield theoretically 1,690 kg of superphosphate, but in practice phosphate rock with an average content of 32 wt.% P_2O_5 (i.e., 70 wt.% BPL) will yield only ca. 1,400 kg of superphosphate.

The industrial, acidulation process, consists first to mix the ground and dried phosphate rock concentrate, usually containing at least 70 wt.% BPL (min. 32 wt.% P_2O_5), with strong concentrated sulfuric acid with a strength ranging from 66 to 73 wt.% H_2SO_4, inside a vertical cone mixer.

Based on the above stoichiometric sulfation reaction, the stoichiometric (theoretical) sulfuric acid-to-solid mass ratio [H_2SO_4 –to-S] is 680 kg (100 wt.% H_2SO_4) per tonne of fluoroapatite.

But in practice, a higher mass ratio of sulfuric acid is always required to offset its consumption by gangue minerals such as fluorite, alumino-silicates, calcite, and iron oxide(s) present in the mined ore. For instance, in the case of a phosphate rock with the chemical composition reported in Table 40, the stoichiometric (theoretical) sulfuric acid-to-solid mass ratio [H_2SO_4–to-S] is 694 kg (100 wt.% H_2SO_4) per tonne of phosphate rock.

Thus in practice, with a 20 wt.% mass excess of sulfuric acid, the practical (actual) sulfuric acid-to-solid mass ratio [H_2SO_4-to-S] becomes 833 kg (100 wt.% H_2SO_4) per tonne of dried phosphate rock. Using commercial sulfuric acid with a strength of 73 wt.% H_2SO_4, this corresponds to a liquid-to-solid mass ratio [A-to-S] ranging from 1,100 to 1,200 kg (73 wt.% H_2SO_4) per tonne prior the water injection.

The key figures related to the sulfation of phosphate rock to produce superphosphate are summarized and reported in Table 40 hereafter.

Table 40 – Sulfation of phosphate rock: key figures

PHOSPHATE ROCK (conc. 70 wt.% BPL)		Sulfation of CaO content	66.67% CaO
Chemical composition		Strength H_2SO_4 (initial)	**73%**
CaO	48.00 wt.%	Strength H_2SO_4 (attack)	64%
P_2O_5	32.00 wt.%	[H_2SO_4-to-Solid](Theor.)	**694** kg/tonne
SiO_2	12.00 wt.%	Mass excess of acid	**20%**
Na_2O	1.60 wt.%	[H_2SO_4-to-Solid](Actual)	**833** kg/tonne
Al_2O_3	1.50 wt.%	[H_2SO_4-to-P_2O_5](Actual)	**2,603** kg/tonne
Fe_2O_3	1.50 wt.%	[Acid-to-Solid](Actual)	**1,141** kg/tonne
K_2O	0.16 wt.%	Solids pulp density	47%
MgO	1.50 wt.%	c_p (solid)	965 J/kg/K
	98.26 wt.%	c_p (acid)	1706 J/kg/K
		c_p (pulp)	**1405** J/kg/K
		$\Delta h_{\text{Sulfation}}$	**-1491** kJ/kg of solid
Mass of solids (dry)	**1000** kg	$\Delta h_{\text{Hydration}}$	-20 kJ/kg of acid
Mass of sulfuric acid	1141 kg	$\Delta h_{\text{Vaporization}}$	2256 kJ/kg of steam
Mass of water (injected)	160 kg	Δh_{Total}	**-198** kJ/kg of mixture
Mass of pulp	**2301** kg	Adiabatic temp rise ΔT	**141 K EXOTHERMIC**

As for the digestion of ilmenite, and titanium slags described in the previous Section 9.1, the sulfation reaction is triggered by the dilution of the concentrated sulfuric acid with a concentration during attack corresponding to a targeted final strength of 64 wt.% H_2SO_4. The excess heat is usually lost by the evolution of steam.

The thick dense slurry which is obtained upon mixing has a pasty consistency with a solid pulp density ranging from 45 to 48 mass percent solids, thus it is feed with a pug mill, and discharged onto a belt conveyor contained inside a heated and air tight enclosure with a residence time of at least one hour.

The hot sulfation cake produced is then cut into individual charges that are further stockpiled and stored for at least 10 days and up to 20 days to ensure the full maturation with the highest P_2O_5 availability.

Afterwards, the matured sulfation cakes are ground, pulverized, and bagged. The noxious emissions from the enclosure containing hydrogen fluoride are scrubbed before being released to the atmosphere.

The schematic flow diagram of the entire process for producing superphosphate from phosphate rock is depicted in Figure 17.

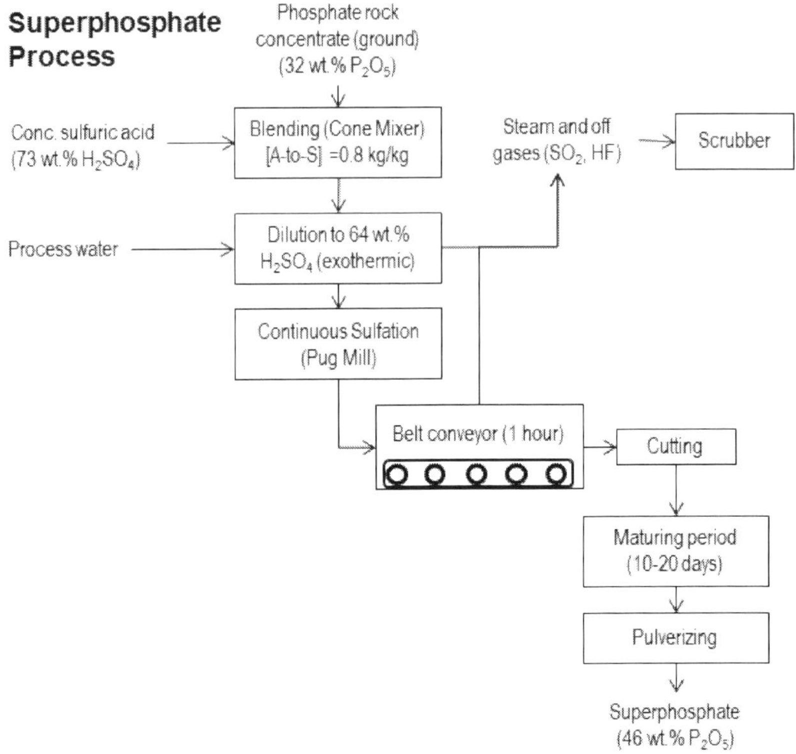

Figure 17 – Flow diagram for the production of superphosphate

Despite being outside the scope of this book, it is interesting to mention that in order to avoid the dilution of the single superphosphate with by-produced gypsum; a smart solution used industrially is to perform the acidulation using phosphoric acid instead of sulfuric acid thus producing the so-called ***triple superphosphate*** with a higher NPK grade of 0-46-0.

$$Ca_{10}(PO_4)_6F_2 + H_3PO_4(l) + 3H_2O(l) = 10CaH_4(PO_4)_2 \cdot H_2O(s) + 2HF(g)$$

Because, the latter being made entirely of monobasic calcium phosphate, it imparts a greater fertilizing value to the final product and thus has a higher saleable market value.

9.3 Phosphoric Acid

Worldwide, there are two main industrial processes used commercially to produce phosphoric acid: (1) the so-called the ***wet process phosphoric acid***

(WPPA) and (2) the dry process. The latter consisting to produce first elementary phosphorus, P_4, by mean of an electric arc furnace, and then phosphoric acid after oxidation of P_4 followed by the hydration of P_2O_5. Therefore, only the former will be described here.

The largest producers of phosphoric acid in the world are in order of annual production capacity [1 Mt = 10^6 tonnes (E)]: China (35 Mt), the United States (14 Mt), Morocco (8.5 Mt), Russia (5 Mt), India (3 Mt), the European Union (2.5 Mt), Saudi Arabia (2 Mt), Brazil (2 Mt), and Jordan (1 Mt).

During the WPPA, **_ortho-phosphoric acid_**, or simply **_phosphoric acid_**, with chemical formula H_3PO_4, CAS No. [7664-38-2], is usually produced by performing the sulfuric acid digestion of finely ground phosphate rock (minus 100 mesh) using 90-95 wt.% H_2SO_4. The overall chemical reaction based on a theoretical fluoroapatite concentrate is as follows:

$$Ca_{10}(PO_4)_6F_2(s) + 10H_2SO_4(l) = 10CaSO_4(s) + 6H_3PO_4(l) + 2HF(g)$$

As described previously for calcium phosphate, the hydrogen fluoride gas, HF, that evolves reacts with the silicates present to form the volatile silicon tetrafluoride that volatilizes according to the chemical reaction below:

$$4HF(g) + SiO_2(s) = SiF_4(g) + 2H_2O(l)$$

Finally, the silicon tetrafluoride is hydrolyzed to yield fluorosilicic acid (FSA) leaving silica as residue (see also Section 9.7) according to:

$$3SiF_4(g) + 2H_2O(l) = 2H_2SiF_6(g) + 3SiO_2(s)$$

The ground phosphate rock assumed to contain at on average 70 wt.% BPL is pulped with strong 93 wt.% concentrated sulfuric acid, with 100 wt.% of the calcia being sulfated, this requires a stoichiometric (theoretical) sulfuric acid-to-solid mass ratio [H_2SO_4-to-S] ranging from 900 up to 1,120 kg (100 wt.% H_2SO_4) per tonne of phosphate rock depending on the chemical composition.

In practice, with a 120 wt.% mass excess, this corresponds to a practical (actual) sulfuric acid-to-solid mass ratio [H_2SO_4-to-S] ranging from 2,100 to 2,500 kg (100 wt.% H_2SO_4) per tonne.

Because of the important amount of heat released, the slurry is cooled and then pumped to a rotary Bird-Prayon tilting pan filtration system that accomplish under vacuum suction the dewatering, the washing, the demoulding of the gypsum cake while recovering an aqueous solution of phosphoric acid having a strength of 50 wt.% H_3PO_4. The latter is further concentrated by performing an energy demanding thermal evaporation in order to yields 70% wt.% H_3PO_4.

The key figures related to the production of phosphoric acid from phosphate rock by the WPPA are summarized and reported in Table 41 hereafter.

Table 41 – Wet phosphoric acid (WPPA) process: key figures

PHOSPHATE ROCK (conc. 70 wt.% BPL)		Sulfation of CaO content	100.00% CaO
Chemical composition		Strength H_2SO_4 (initial)	**93%**
CaO	48.00 wt.%	Strength H_2SO_4 (attack)	93%
P_2O_5	32.00 wt.%	[H_2SO_4-to-Solid](Theor.)	**974** kg/tonne
SiO_2	12.00 wt.%	Mass excess of acid	**120%**
Na_2O	1.60 wt.%	[H_2SO_4-to-Solid](Actual)	**2,143** kg/tonne
Al_2O_3	1.50 wt.%	[H_2SO_4-to-P_2O_5](Actual)	**6,696** kg/tonne
Fe_2O_3	1.50 wt.%	[Acid-to-Solid](Actual)	**2,304** kg/tonne
K_2O	0.16 wt.%	Solids pulp density	30%
MgO	1.50 wt.%	c_p (solid)	965 J/kg/K
	98.26 wt.%	c_p (acid)	1467 J/kg/K
		c_p (pulp)	**1268** J/kg/K
		Δh Sulfation	**-2307** kJ/kg of solid
Mass of solids (dry)	**1000** kg	Δh Hydration	0 kJ/kg of acid
Mass of sulfuric acid	2304 kg	Δh Vaporization	2256 kJ/kg of steam
Mass of water (injected)	0 kg	Δh Total	**-588** kJ/kg of mixture
Mass of pulp	**3304** kg	Adiabatic temp rise ΔT	**464 K EXOTHERMIC**

Some phosphoric acid originating from the slurry cooler is also added to the slurry. Inside the digester, the mixture is vigorously agitated while the operating temperature is kept between 85-100°C.

The acidulation process produces a thick suspension with a pulp density of ca. 30 mass percent solids essentially made of precipitated calcium sulfate(s) as by-products together with unreacted coarser grains of the original phosphate rock. The conversion efficiency of the sulfuric acid digestion is usually excellent 95-98%.

The schematic flow diagram of the entire WPPA process from phosphate rock is depicted in Figure 18.

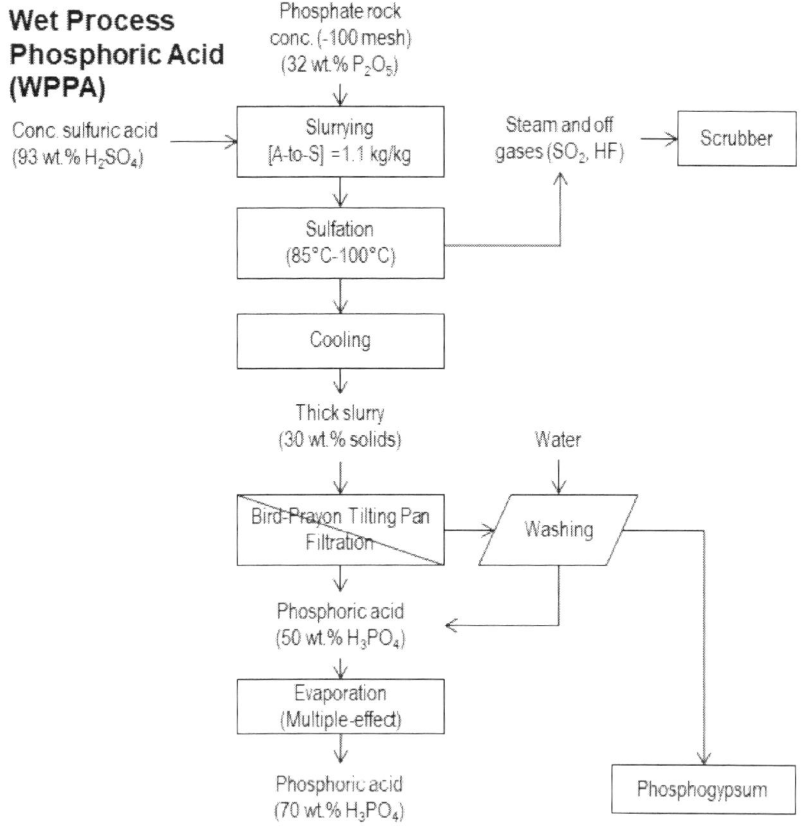

Figure 18 – Flow diagram for the production of wet phosphoric acid

Depending on the operating temperature, the P_2O_5 content, and the water content of the pregnant rich solution, either gypsum ($CaSO_4.2H_2O$), *calcium sulfate hemihydrate* ($CaSO_4.0.5H_2O$) or even *anhydrite* ($CaSO_4$) or there mixtures are by-produced.

Gypsum that was once the only by-product from the WPPA, is still today the most common by-product as the related process can address a plethora of phosphate rocks exhibiting various chemistries.

However, following the first oil crisis in 1973, when the skyrocketing energy prices imposed to modify the traditional process by minimizing the consumption of energy and a lower usage of steam, this approach resulted in a revamped and modernized WPPA by-producing in addition to gypsum, the calcium sulfate hemihydrate, or a mixture of gypsum and the calcium sulfate hemihydrate.

Moreover, when the calcium sulfate hemihydrate is generated as by-product, it allows the WPPA to produce phosphoric acid solutions with higher strengths without requiring further concentration with an energy demanding thermal evaporation step and with a significant reduction in consumption of process water.

9.4   Lithium from Spodumene

Winning lithium from silicate minerals such as **spodumene** ($LiAlSi_2O_6$) or in a lesser extent the mica **lepidolite** $[K_2(Li,Al)_3(Si_3AlO_{10})(F,OH)_2]$ **petalite** ($Li_2Al_2Si_8O_{20}$), and formerly **zinnwaldite** $[KLiFeAl(Si_3AlO_{10})(OH,F)_2]$ is always a highly energy demanding chemical process with the chemical plant usually located near the hard rock pegmatite type deposit. Before describing the acid-roast process of spodumene, it is worth to summarize a brief history of the production of lithium from spodumene.

In Central Europe, from 1890 until 1945, the lithiferous mica zinnwaldite was by-produced from the mining of wolframite and cassitcrite from the Zinnwald-Georgenfeld cross-border tin-tungsten (Sn-W) deposit in the ErzGebirge region near Cinovec (formerly Zinnwald) by the *Stahlwerk Becker* AG [81].

From 1923 until the end of World War II, the zinnwaldite mica concentrate was shipped to the Hans-Heinrich Werke chemical facilities in Langelsheim near Frankfurt-am-Main, Germany, owned by the *Metallgesellsfhaft* AG and then roasted with limestone and gypsum to produce lithium fluoride (LiF) and potassium sulfate (K_2SO_4). At that time, *Metallgesellsfhaft* was the first and only company to implement such process worldwide [82].

During the same period in the USA, the *Maywood Chemical Works* started in 1929 the production of lithium carbonate from the limestone-roasting of spodumene at its plant located in Maywood, New Jersey, beneficiating from the technical support and transfer of technology from *Metallgesellsfhaft* AG. The spodumene was mined and beneficiated from hard rock pegmatite veins located in the Black Hills, South Dakota.

In was only in the 1940, that the sulfation process or **acid-roast process** was invented by Ellestad and Leute [83] from the *Metalloy Corporation* and implemented industrially in chemical processing facilities established in Bessemer City, North Carolina. The company was later renamed the *Lithium Corporation of*

[81] DEBERITZ, J. (1993) *Lithium: Production and Application of a Fascinating and Versatile Element.* Die Bibliotehk des Wissenschaft, Vol.2 Verlag Modern Industrie AG & Co., Landsberg, Germany.

[82] MOTOCK, G.T. (1945) *Lithium extraction and uses.* Field Information Agency, Technical (FIAT) Field Report No. 295, Joint Intelligence Objectives Agency, Washington DC, London, UK.

[83] ELLESTAD, R.B.; and LEUTE, K.M. (1950) Method of Extracting Lithium Values from Spodumene Ores. U.S. Patent Application 2,516,109 (Metalloy Corporation), July 25th, 1950.

America (LithCoA) and became a subsidiary of the *Foote Machinery Company* (FMC). The spodumene was mined locally from an open pit in the lithium tin-spodumene belt near Charlotte, North Carolina, USA where large spodumene ore deposits were found (e.g., Cherryville, and Kings Mountain).

As a consequence, at the end of World War II, *Metallgesellsfhaft* abandoned the limestone-gypsum roasting of zinnwaldite in favor to the novel "American route" for the acid-roasting of spodumene that was carried on until the 1970s. The concentrate was then entirely imported from the US Company *Foote Mineral Co.* (later *Cyprus Foote AMAX*) and mined at Kings Mountain, North Carolina. *Metallgesellsfhaft* AG who later acquired *Cyprus Foote* in 1999 was renamed *Chemetall Foote*, and since 2015 it is owned by *Albemarle*.

Similarly *FMC Lithium* continued the processing of spodumene from North Carolina and even during a short period from 1958 to 1964, spodumene was imported from the Canadian company *Quebec Lithium*. Nowadays *FMC Lithium* changed its name to *Livent*.

However, in 1999, due to the economic downturn, the spodumene prices at their lowest, and the rapid expansion of the lithium extraction from brine fields (i.e., Salars) in South America, the mining, and the beneficiation of spodumene was discontinued in the United States and the existing sulfation plants dismantled.

It is only recently that the US public company *Piedmont Lithium* announced its intention to reopen an open pit mine and to process spodumene domestically to produce again lithium carbonate and lithium hydroxide in the US.

In Canada, two projects for mining and performing the acid-roasting of spodumene arose in the 2010s, the first lead by the resurrected company named *Quebec Lithium* focussed on the Lacorne hard rock pegmatite deposit located in the Abitibi region, 60 km north of Val d'Or, Quebec, while the public company *Nemaska Lithium* was aiming to mine spodumene concentrate from the Wabouchi hard rock pegmatite deposit, 280 kilometres NNE of Chibougamau, Quebec. The latter was aiming to perform the acid-roasting, and to convert the lithium sulfate brine by electro-dialysis into lithium hydroxide. Both companies went bankrupt in 2016 and 2020 respectively.

Since then *Quebec Lithium* was renamed *North American Lithium Inc.* (NAL) and acquired by a Chinese consortium that restarted for a short period the production of spodumene concentrate. In 2019, production was halted again, then in 2021 the company was officially acquired by *Sayona* (75%) and *Piedmont Lithium* (25%).

On the other hand, *Nemaska Lithium* is now a private company owned by a consortium consisting of *The Pallinghurst Group* from the UK, *Orion Mine Finance* from the USA, and the *Gouvernement du Québec* through its financial arm *Investissement Québec*.

Globally, nowadays the industrial acid-roast process is only performed industrially in Western Australia, and China with the major producers listed in Table 42.

Table 42 – Major lithium producers processing spodumene by sulfation

Major spodumene producers and processors	Spodumene deposit and mine location
Albemarle (previously Cyprus Foote AMAX, Chemetall Foote, Rockwood Lithium Holdings)	Greebushes, Western Australia (49% stake) Wodgina, Western Australia
Livent (formerly FMC Lithium, Lithium Corporation of America, Metalloy)	Bessemer City, North Carolina (closed in 1999)
Tianqui Lithium	Greebushes, Western Australia (51% stake)
Jiangxi Ganfeng Lithium Co. Ltd.	Mount Marion, Western Australia
Pilbara Minerals	Pilgangoora Western Australia Pilgan plant Ngungaju plant

The schematic flow diagram describing the Ellestar acid-roast process of spodumene is depicted in Figure 19 and summarized briefly hereafter.

Prior performing the sulfation roasting, the spodumene concentrate (4.5 to 7.5 wt.% Li_2O) is obtained by beneficiation techniques such as flotation and it must be thermally processed in order to convert the naturally occurring mineral phase alpha-spodumene (monoclinic) to the tetragonal beta-spodumene [84].

Due to the irreversible phase transition accompanied by an important volume change (+8 vol.%)[85], the grains decrepitate forming a network of cracks within the lattice structure with high specific area thus enhancing its chemical reactivity towards sulfuric acid.

For that reason, only beta-spodumene is amenable to be sulfated with high conversion efficiencies usually ranging from 95% to 99%. Theoretically, the temperature of the irreversible phase transition from alpha to beta phase is 1075°C

[84] CARDARELLI, F. (2018) *Lithium* in *Materials Handbook. A Concise Desktop Reference, Third edition.* Springer, Cham, London, New York, pp.318-340.

[85] DESSEMOND, C.; SOUCY, G.; HARVEY, J.P.; and OUZILLEAU, P. (2020) Phase Transitions in the α–γ–β spodumene thermodynamic system and impact of γ-spodumene on the efficiency of lithium extraction by acid leaching. *Minerals* **10**(519)1-23.

but in practice little departure ranging from 1075°C and 1100°C was observed among various spodumene feeds.

The decrepitating process is performed historically inside brick line rotary kilns and more recently utilizing gas suspension shaft kilns or fluidized bed roasters. In both equipment's, the efficient conversion depends on the calcination temperature, the particle size distribution of the spodumene concentrate, and the residence time. Usually, coarser spodumene concentrate produced via dense media separation with feed size ranging from 3 mm up to 12 mm must be processed using rotary kilns with residence time of nearly half an hour, while shaft kilns operating under pneumatic conveying conditions require finer material below 1 mm that offers enhanced flow ability with residence time less than a minute.

However, owing to the flux properties of remaining gangue silicate minerals, the temperature during roasting must be carefully controlled and kept always below 1400°C to prevent the formation of eutectics between alpha-spodumene and other silicate minerals. After decrepitating and cooling, the beta-spodumene is softly ground in a rubber lined ball-mill down to 150 μm (-100 mesh) particle size in order to enhanced the specific area.

The sulfuric acid roasting consists to mix the finely ground and decrepitated beta-spodumene with concentrated sulfuric acid (93 wt.% H_2SO_4) to obtain a paste with 35 wt.% excess of sulfuric acid.

If all the lithium ad aluminum values are extracted, the theoretical acid number [H_2SO_4-to-S] is 930 kg (100 wt.% H_2SO_4) per tonne of spodumene. Therefore, it corresponds to a practical (actual) sulfuric acid number [A-to-S] of 1,350 kg per tonne of spodumene. The thick suspension exhibits a pulp density of 45 mass percent solids. The key figures related to the sulfation of spodumene according to this theoretical scenario are summarized and reported in Table 43.

However, industrially only the lithium values are extracted while leaving behind an alumino-silicate solid. In this case, the stoichiometric (theoretical) sulfuric acid-to-solid mass ratio [H_2SO_4-to-S] is reduced to 235 kg (100 wt.% H_2SO_4) per tonne, and the practical (actual) sulfuric acid-to-solid mass ratio [H_2SO_4-to-S] becomes 341 kg (100 wt.% H_2SO_4) per tonne of spodumene concentrate. The pasty mass formed has a pulp density of 75 mass percent solids. The key figures related to this second scenario are summarized and reported in Table 44 hereafter.

Table 43 – Sulfation of spodumene: key figures (lithium and aluminum values)

SPODUMENE (concentrate)			
Chemical composition		Strength H_2SO_4 (initial)	**93%**
SiO_2	67.63 wt.%	Strength H_2SO_4 (attack)	90%
Al_2O_3	24.10 wt.%	[H_2SO_4-to-Solid](Theor.)	**930** kg/tonne
Li_2O	5.98 wt.%	Mass excess of acid	**35%**
Na_2O	0.53 wt.%	[H_2SO_4-to-Solid](Actual)	**1,256** kg/tonne
K_2O	0.50 wt.%	[Acid-to-Solid](Actual)	**1,350** kg/tonne
Fe_2O_3	0.76 wt.%	Solids pulp density	43%
CaO	0.20 wt.%	c_p (solid)	828 J/kg/K
MgO	0.30 wt.%	c_p (acid)	1467 J/kg/K
	100.00 wt.%	c_p (pulp)	**1213** J/kg/K
		$\Delta h_{Sulfation}$	**-645** kJ/kg of solid
Mass of solids (dry)	**1000** kg	$\Delta h_{Hydration}$	-49 kJ/kg of acid
Mass of sulfuric acid	1350 kg	$\Delta h_{Vaporization}$	2256 kJ/kg of steam
Mass of water (injected)	45 kg	Δh_{Total}	**-166** kJ/kg of mixture
Mass of pulp	**2395** kg	Adiabatic temp rise ΔT	**137 K EXOTHERMIC**

Table 44 – Sulfation of spodumene: key figures (lithium value only)

SPODUMENE (concentrate)			
Chemical composition		Strength H_2SO_4 (initial)	**93%**
SiO_2	67.63 wt.%	Strength H_2SO_4 (attack)	93%
Al_2O_3	24.10 wt.%	[H_2SO_4-to-Solid](Theor.)	**235** kg/tonne
Li_2O	5.98 wt.%	Mass excess of acid	**35%**
Na_2O	0.53 wt.%	[H_2SO_4-to-Solid](Actual)	**317** kg/tonne
K_2O	0.50 wt.%	[Acid-to-Solid](Actual)	**341** kg/tonne
Fe_2O_3	0.76 wt.%	Solids pulp density	75%
CaO	0.20 wt.%	c_p (solid)	828 J/kg/K
MgO	0.30 wt.%	c_p (acid)	1467 J/kg/K
	100.00 wt.%	c_p (pulp)	**973** J/kg/K
		$\Delta h_{Sulfation}$	**-485** kJ/kg of solid
Mass of solids (dry)	**1000** kg	$\Delta h_{Hydration}$	0 kJ/kg of acid
Mass of sulfuric acid	341 kg	$\Delta h_{Vaporization}$	2256 kJ/kg of steam
Mass of water (injected)	0 kg	Δh_{Total}	**-322** kJ/kg of mixture
Mass of pulp	**1341** kg	Adiabatic ΔT	**331 K EXOTHERMIC**

The pasty mass is then roasted inside a small kiln up to 250°C to 300°C with a residence time of 20 minutes in order to convert the lithium values into water soluble lithium sulfate, Li_2SO_4. The clinker obtained is then quenched into cold water, allowing leaching out the lithium sulfate and other soluble metal sulfates. The leaching stage is performed during 2 hours.

The pregnant leach solution (PLS) obtained is an impure solution of lithium sulfate containing traces of iron, aluminum, and other alkali metals cations (e.g., Ca^{2+}, and Mg^{2+}). The excess of sulfuric acid is then neutralized at pH 6.5 by adding powdered natural calcium carbonate (i.e., ground limestone). The impurities that form insoluble precipitates yield a s [e.g., $CaSO_4.2H_2O$, $FeCO_3$, and $Al(OH)_3$] from which the solids are removed either by filtration or centrifugation and the wet sludge disposed-off and landfilled yielding a purified solution of lithium sulfate.

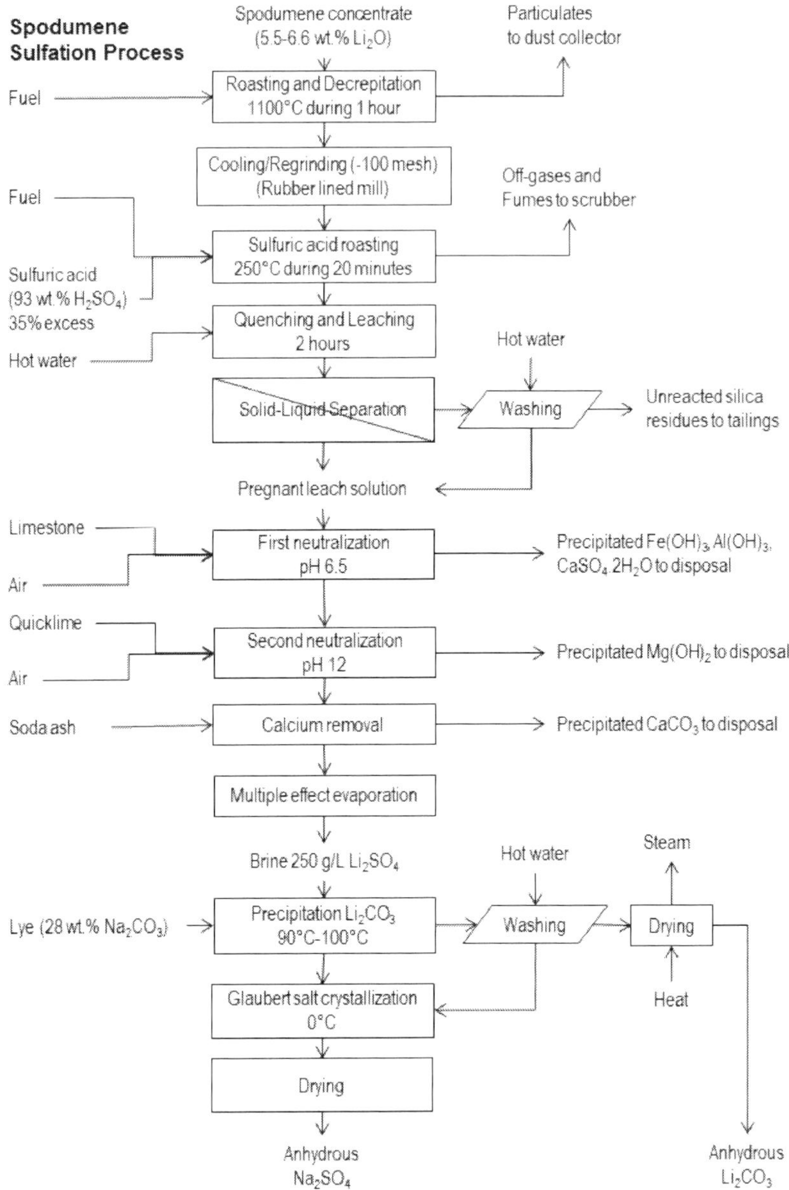

Figure 19 – Flow diagram for the production of lithium from spodumene

Then, the clear liquor is treated with slacked lime (i.e., calcium hydroxide, $Ca(OH)_2$), until a pH of 12 is reached in order to precipitate all the magnesium values as magnesium hydroxide, $Mg(OH)_2$.

Finally, anhydrous soda ash (i.e., sodium carbonate, Na_2CO_3) is added to precipitate the remaining traces of calcium and magnesium, which are removed later by filtration.

Afterwards, the solution is pH-adjusted between 7 and 8 by adding concentrated sulfuric acid and then concentrated by evaporation using a multiple effect evaporator to concentrated liquor containing between 200 and 250 g/L of lithium sulfate.

Then, the slightly soluble lithium carbonate, Li_2CO_3 is precipitated at 90-100°C by metering the addition of a dilute lye of sodium carbonate (28 wt.% Na_2CO_3), and then separated by centrifugation, washed, dried, and bagged for sale "as is" or used as intermediate chemical.

The remaining mother liquor still contains about 15 mass percent of lithium with large amount of sodium sulfate. Thus, it requires the removal of sodium as sodium sulfate decahydrate, $Na_2SO_4.10H_2O$ (i.e., Glauber salt's) by crystallization at 0°C. The wet crystals are further dried, ground, and the anhydrous sodium sulfate powder bagged and ready to be used in the detergent and glass making industry (see Section 11.3).

9.5   Aluminum Sulfate

Aluminium sulfate also called ***alum*** in the trade, with chemical formula $Al_2(SO_4)_3.18H_2O$, CAS No. [10043-01-3], is the second most important aluminum chemical produced worldwide just after alumina. In 2020, the major aluminium sulfate producing countries are in order of importance: China, Turkey, Sweden, Germany, and Indonesia.

Actually, aluminium sulfate is primarily used as flocculating agent for potable water treatment and wastewater treatment. Actually, because the aluminium cation, Al^{3+}, exhibits a pronounced amphoteric behavior depending on the pH (i.e., Al^{3+}, $Al(OH)_4^-$) together with a strong polarizing ability due to a high charge together with small ionic radius, which makes it an excellent coagulant for clarification of negatively charged colloidal particles such as clays, colloid silica, or suspended organic matter acting as flocculating agent and also performing the removal of phosphorus. For that purpose, the chemical is supplied either as the solid hydrated salt or as a saturated aqueous solution containing 48 wt.% $Al_2(SO_4)_3$.

The second major industrial application is in the pulp and paper industry where it is used in paper mills for charge neutralization, rosin sizing, and pitch control. Finally, minor utilizations are in the textile industry for fixing dyes to fabrics and in synthetic catalyst production, and in farming as a litter amendment for ammonia control.

Aluminum sulfate octadecahydrate is usually prepared industrially from the sulfation of two different raw materials both originating directly or indirectly from bauxite: (1) by the so-called *Giulini process* that consists to perform the sulfuric acid digestion under pressure of *aluminum trihydrate* (ATH) originating from the caustic digestion of bauxite by the Bayer process; and (2) by the sulfuric acid digestion of low grade bauxite ores.

9.5.1   The Giulini Process

The former route is used essentially when a high purity chemical is required. The *Giulini process* was invented by Rüter et al., and patented in 1965 by the German company *Gebruder Giulini GmbH* [86] an independent and non-integrated producer of alumina with a plant in Ludwigshafen at that time. It was described in details by Helmboldt [87]. A schematic flow diagram of the Giulini process is depicted in Figure 20.

[86] RÜTER, H.; CHERDRON, E.; and FÄSSLE, F. (1965) *Process for the production of aluminum sulfate melt*. U.S. Patent 3,226,188 (Gebruder Giulini GmbH), December 28th, 1965.

[87] HELMBOLDT, O. (1997) *Aluminum Sulfate*. in HABASHI, F. (ed) (1997) *Handbook of Extractive Metallurgy*. Volume II, Wiley-VCH, Weinheim, Germany, pp. 1103-1104.

Giulini Process

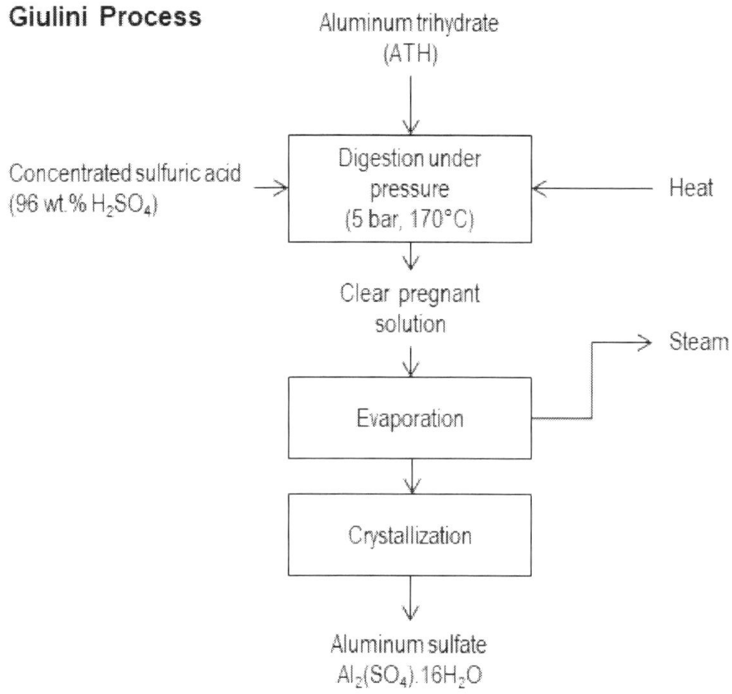

Figure 20 – Flow diagram for the production of alum by the Giulini process

During the above process, the aluminum trihydrate (ATH) is digested with concentrated sulfuric acid (66 wt.% H_2SO_4) inside an autoclave at 170°C and under a pressure of 500 kPa. Aluminium sulfate is produced according to the following exothermic chemical reaction:

$$2Al(OH)_3 + 3H_2SO_4 + 8H_2O \rightarrow Al_2(SO_4)_3.14H_2O$$

Afterwards, the concentrated liquor produced undergoes evaporation from which crystals of aluminum sulfate are produced, hammer-milled down to 200 mesh (75 μm) and bagged.

Selected major producers using the Giulini process worldwide are reported in Table 45.

Table 45 – Major aluminum sulfate producers from ATH

Geographical area	Producers
United States	Affinity, Chemicals, Dallas. C & S Chemical Inc. General Alum & Chemical Corp. (GAC). Universal USALCO
Europe	Kemira, Finland Brenntag, Sweden
Asia	Nippon Light Metals, Japan Nankai Chemicals Co., Ltd., Japan Shandong Sanfeng Group Co., Ltd., China

9.5.2 Aluminum Sulfate from Bauxite

This route is essentially used where low grade bauxite ores are available with a high silica and low iron content making them unsuitable to be processed by the Bayer process. The treatment produces though impure aluminum sulfate liquors that are only suitable to be used as flocculating agent for wastewater treatment.

It is currently performed commercially in the United States, the United Kingdom, Spain, Sweden, India, Russia, Turkey, and Venezuela (see Table 46).

Table 46 – Major aluminum sulfate producers from bauxite

Geographical area	Producers
United States	Chemtrade Logistics Inc. Univar Solutions
Europe	Industrial Chemicals Ltd., UK Spain Sweden Turkey
India	Vedanta
Russia	State owned
Venezuela	State owned

Several grades of ground bauxite are specifically produced for the manufacture of aluminum chemicals. Usually, these grades are all characterized by low content of iron (III) and the available alumina should be readily soluble in sulfuric acid.

For instance, the specifications of chemical grades of bauxite must be within the ranges reported in Table 47.

Table 47 – Specifications for chemical grade bauxite

Oxide	Mass percentage (wt.%)
Al_2O_3	56.5 – 60.5
SiO_2	4.25 – 9.0
TiO_2	2.25 – 3.5
Fe_2O_3	1.5 – 3.0
LOI	28.0 – 31.0

The schematic flow diagram of the entire process is depicted in Figure 21.

The ground bauxite with a particle size of 80 mass percent passing 200-mesh (75 µm) is mixed with 60°Be sulfuric acid (ca. 78 wt.% H_2SO_4) inside lead-lined steel digesters, where the reactants are thoroughly mixed with impellers and heated using superheated steam due to the endothermic sulfation reaction.

Depending on the initial chemical composition of the bauxite, the stoichiometric (theoretical) sulfuric acid-to-solid mass ratio [H_2SO_4-to-S] ranges from 1,700 to 1,900 kg per tonne. Based on the above sulfuric acid strength of 78 wt.% and with no excess acid the practical acid number [A-to-S] ranges from 2,175 to 2,436 kg per tonne of bauxite forming a slurry with a pulp density of 30 mass percent solids.

The key figures related to the sulfation of bauxite to produce alum are summarized and reported in Table 48 hereafter.

Table 48 – Sulfation of bauxite: key figures

BAUXITE (conc. 61 wt.% Al2O3)			
Chemical composition		Strength H_2SO_4 (initial)	**78%**
Al_2O_3	61.30 wt.%	Strength H_2SO_4 (attack)	78%
SiO_2	4.25 wt.%	[H_2SO_4-to-Solid](Theor.)	**1,867** kg/tonne
TiO_2	3.45 wt.%	Mass excess of acid	**0%**
Fe_2O_3	3.00 wt.%	[H_2SO_4-to-Solid](Actual)	**1,867** kg/tonne
H_2O	28.00 wt.%	[Acid-to-Solid](Actual)	**2,393** kg/tonne
	100.00 wt.%	Solids pulp density	29%
		c_p (solid)	1762 J/kg/K
		c_p (acid)	1639 J/kg/K
		c_p (pulp)	**1506** J/kg/K
		Δh Sulfation	**-489** kJ/kg of solid
Mass of solids (dry)	**1000** kg	Δh Hydration	0 kJ/kg of acid
Mass of sulfuric acid	2393 kg	Δh Vaporization	2256 kJ/kg of steam
Mass of water (injected)	0 kg	Δh Total	**392** kJ/kg of mixture
Mass of pulp	**3393** kg	Adiabatic temp rise ΔT	**-260 K ENDOTHERMIC**

Upon completion of the sulfation reaction, the sulfated mass is then dissolved with acidic water and all the ferric iron is reduced chemically to ferrous iron *in-situ* by simply adding barium sulfide (i.e., **black ash** with 70 wt.% BaS) obtained by calcining barite ($BaSO_4$) with a source of carbon (e.g., anthracite coal) inside a rotary kiln.

The slurry from the reactors is then sent through a series of thickeners, operated counter-currently, in order to remove unreacted solids and the precipitated barium sulfate ($BaSO_4$) that are thoroughly water washed prior to be disposed as wet sludge.

The clarified pregnant solution containing all the aluminum sulfate is concentrated in an open, steam-coil heated kettle from 26 wt.% to 48 wt.% $Al_2(SO_4)_3$. The concentrated liquor is poured into flat pans, where it is cooled and completely solidified. The solid cake is then broken and ground to suitable size and bagged for shipping.

For most water treatment plants, the last operation unit for concentrating the liquor becomes unnecessary expense; therefore the concentrated 48 wt.% aluminum sulfate solution is used directly instead. Conversely, the addition of sodium sulfate as a by-product allows the precipitation of the so-called **Na-alum** with the chemical formula $NaAl(SO_4)_2.12H_2O$.

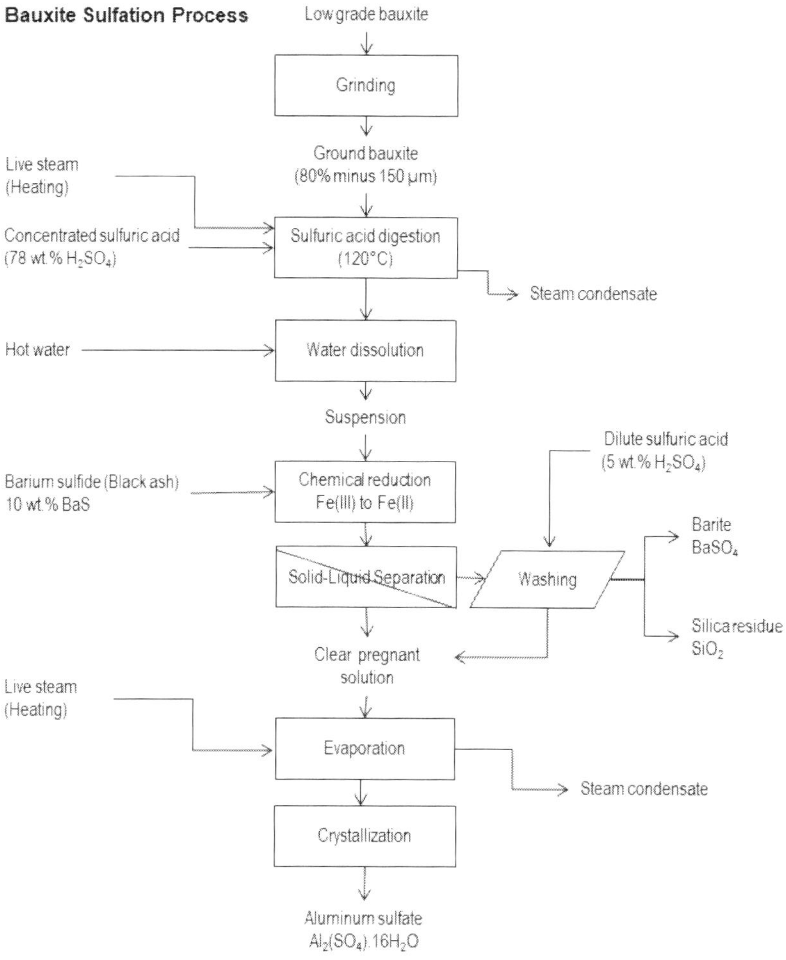

Figure 21 – Flow diagram for the production of aluminum sulfate from bauxite

9.6 Hydrofluoric Acid from Fluorspar

Hydrofluoric acid, is an aqueous solution of hydrogen fluoride with the chemical formula, HF, CAS No. [7664-39-3], and it is the most important fluorine chemical with ca. 1.7 million tonnes of anhydrous HF equivalent produced worldwide in 2021. China is the largest producing country by far followed by Mexico, Germany, Japan, Spain, and the USA these six countries control the global hydrofluoric acid market.

It is used "as is" in the pickling of stainless steels and non-ferrous metals, glass etching, and oil refining (alkylation) or as an intermediate product for the manufacture of fluorine inorganic compounds such as: aluminum fluoride, AlF_3, CAS No. [7784-18-1], and synthetic **cryolite**, Na_3AlF_6, CAS No. [15096-52-3], both used during the molten fluoride electrolysis of alumina, fluorine gas, and uranium hexafluoride, UF_6, CAS No. [7783-81-5], the latter used for the isotopic enrichment of 235-uranium during the nuclear fuel cycle; or organic fluorine compounds such as fluorocarbons, and fluorinated polymers (e.g., PTFE, PFA, PVDF).

The key figures related to the sulfation of fluorspar to produce hydrofluoric acid are summarized and reported in Table 49 hereafter.

Table 49 – Sulfation of fluorspar: key figures

FLUORSPAR (acid grade: 97 wt.% CaF2)		
Chemical composition	Strength H_2SO_4 (initial)	**93%**
CaF2 97.00 wt.%	Strength H_2SO_4 (attack)	93%
	[H_2SO_4-to-Solid](Theor.)	**1,219** kg/tonne
	Mass excess of acid	**0%**
	[H_2SO_4-to-Solid](Actual)	**1,219** kg/tonne
	[Acid-to-Solid](Actual)	**1,310** kg/tonne
	Solids pulp density	43%
	c_p (solid)	716 J/kg/K
	c_p (acid)	1467 J/kg/K
	c_p (pulp)	**1103** J/kg/K
	Δh Sulfation	**41** kJ/kg of solid
Mass of solids (dry) **1000** kg	Δh Hydration	0 kJ/kg of acid
Mass of sulfuric acid 1310 kg	Δh Vaporization	2256 kJ/kg of steam
Mass of water (injected) 0 kg	Δh Total	**107** kJ/kg of mixture
Mass of pulp **2310** kg	Adiabatic temp rise ΔT	**-97 K ENDOTHERMIC**

The mineral **fluorite** (CaF_2) also called **fluorspar** in the trade is the major raw material for the production of **hydrofluoric acid** (HF). The **acid grade** fluorspar or **acid-spar** used for the production of HF requires a content of at least 97 wt.% CaF_2 and a low silica content.

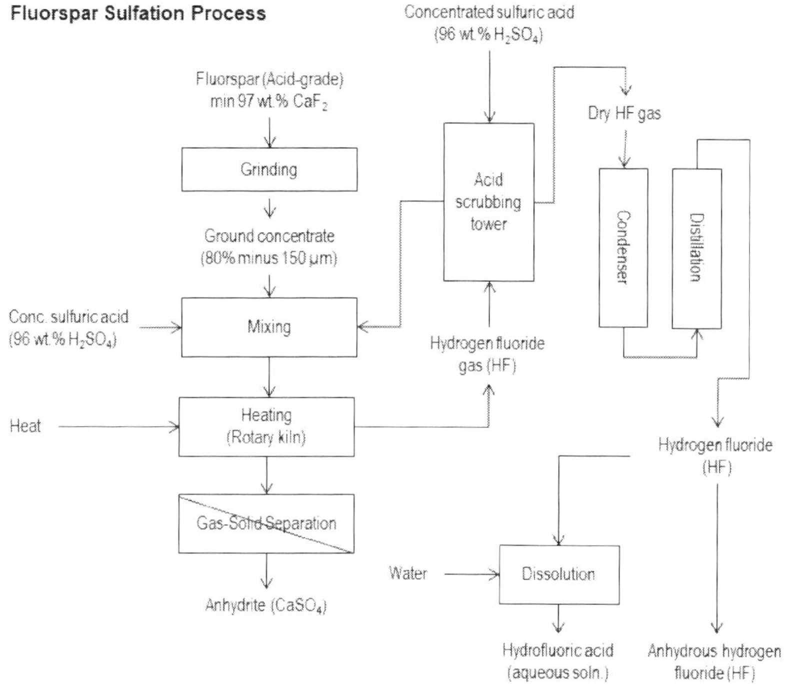

Figure 22 – Flow diagram for the production of hydrofluoric acid from fluorspar

During the industrial process with the schematic flow diagram depicted in Figure 22, the hydrogen fluoride gas is obtained following a two-stage process consisting to pre-mix inside a reactor the ground and dry fluorite concentrate with cold concentrated sulfuric acid (93 wt.% H_2SO_4) with a stoichiometric (theoretical) sulfuric acid-to-solid mass ration [H_2SO_4-to-S] of 1,219 kg (100 wt.% H_2SO_4) per tonne of fluorspar. Because the chemical sulfation reaction is endothermic, the resulting slurry is feed directly into an externally heated rotary kiln. The fluorite is then decomposed according to the following chemical reaction:

$$CaF_2(s) + H_2SO_4(l) = CaSO_4(s) + HF(g)$$

The hydrogen fluoride gas is scrubbed counter currently inside tall tower with concentrated sulfuric acid, condensed and distilled to yield anhydrous hydrofluoric acid (AHF). The solid anhydrous calcium sulfate (anhydrite) left behind is simply disposed-off and landfilled.

 – Major producers of hydrofluoric acid from fluorspar

Company	Location
Derivados del Fluor SA	Ontón (Spain)
Fluorchemie GmbH	Dohna (Germany) Stulln (Germany)
Fluorsid S.p.A.	Assemini (Italy) Porto Marghera (Italy)
Honeywell Specialty Chemicals GmbH	Seelze (Germany)
Lanxess Deutschland GmbH	Leverkusen (Germany)
Solvay Fluor GmbH	Bad Wimpfen (Germany)
Do-Fluoride Dongyue Group Daikin, Sinochem Yingpeng Chemical Fluoride Chemicals	Various locations in China
Honeywell Specialty Chemicals DuPont Alcoa	USA
Honeywell	Canada
QF IQM Quimibasicos	Mexico

It is important to note that pure nickel grade 200, Monel® 400, carbon, magnesium, and lead metals are chemically resistant to concentrated HF and above 60 wt.% HF steel becomes a satisfactory corrosion resistant material.

9.7 Hydrofluoric Acid from Fluorosilicic Acid

As already reported in the Section 9.2 dealing with the phosphate fertilizers industry, a significant amount of *fluorosilicic acid* or *fluosilicic acid* (H_2SiF_6) CAS No. [16961-83-4], sometimes called *hexafluorosilicic acid*, denoted by the acronym FSA and HFS in the trade, is by-produced during the production of phosphate by acidulation.

Actually, most phosphate rock contains between 2.5 wt.% and 4.4 wt.% fluorine mostly hosted in the fluoroapatite mineral phase. Therefore, it represents now an alternative source for the production of HF. The process was originally devised by the Polish company *Zaklady Chemiczne LUBON SA* and consists to decompose an aqueous solution of FSA with strong sulfuric acid into silicon tetrafluoride (SiF_4) and gaseous HF. The latter leaves the digestion vessel first and is evaporated further to yield anhydrous HF. Afterwards, the SiF_4 is released by further heating and it is absorbed into the upstream feed solution to generate additional FSA and leaving the excess silicon as pure silica co-product. The major producers of fluorosilicic acid are reported in Table 51.

Table 51 – Major fluorosilicic (fluosilicic) acid producers

Company	Location
Gelest Inc. (Mitsubishi Chemical)	Morrisville, PA, USA
Hawkins	Roseville, MN, USA
Honeywell International Inc.	USA
Hydrite Chemical	Brookfield, WI, USA
IXOM Operations Pty Ltd.	Mt Maunganui South, New Zealand
KC Industries, LLC	Mullberry, FL, USA
Merck KGaA	Darmstad, Germany
Napco Chemical Company	Spring, TX, USA
Solvay America LLC.	USA
Univar Solutions Inc.	Downers Grove, IL, USA
Xinxiang Yellow River Fine Chemical Industry Co., Ltd	Xinxiang, Hunan, China

9.8 The Brush Beryllium Process

The sulfation process of beryl a cyclosilicate mineral with the chemical formula, $Be_3Al_2Si_6O_{18}$, is used to produce beryllium and aluminum sulfates. It was invented in 1935 by Sawyer and Kjellgreen [88] and implemented industrially by the *Brush Beryllium Co.* (later *Brush-Wellman, Inc.* nowadays *Materion*) in Cleveland, Ohio. A schematic flow diagram of the Brush Beryllium process is depicted in Figure 23.

The industrial process consists to perform the crushing of a coarse beryl concentrate. The crushed beryl is then melted inside an arc furnace at 1600°C-1625°C and the molten bath quenched by simply pouring the melt directly into cold running water to yield beads of a reactive glassy material.

---

[88] SAWYER, C.B.; and KJELLGREN, B. (1935) *Process of obtaining beryllium and aluminum compounds.* U.S. Patent 2,018,473 (Brush Beryllium).

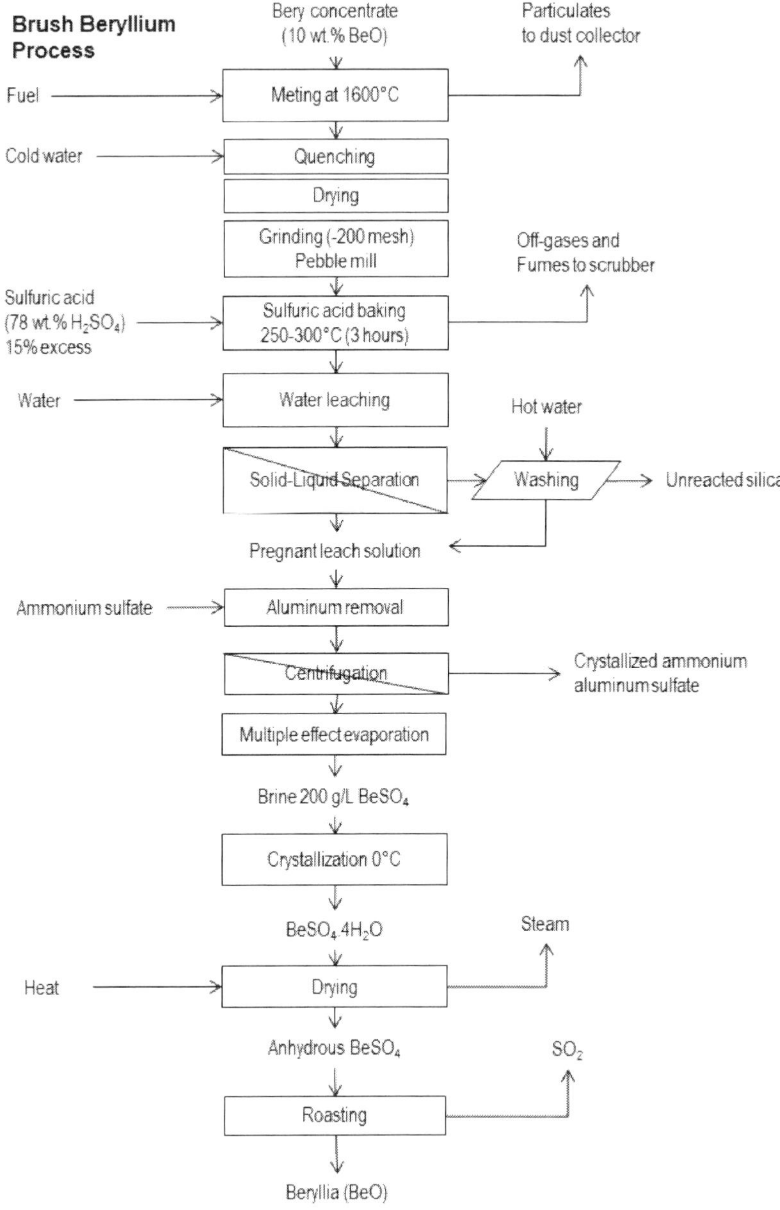

Figure 23 – Flow diagram for the Brush Beryllium process

This mandatory operation is intended to break down the crystal structure and convert the rather chemically inert beryl into a more acid soluble glass. This approach resembles to some extent to the decrepitating of spodumene as described in Section 9.4.

After drying, the glass beads are ground inside a pebble mill lined with impact resistant mullite bricks down to a particle size of minus 75 μm (200 mesh Tyler). The ground beryl flour is then thoroughly mixed with 60°Be sulfuric acid (77.7 wt.% H_2SO_4). The stoichiometric (theoretical) sulfuric acid-to-mass ratio [H_2SO_4-to-S] is 1,087 kg (100 wt.% H_2SO_4) per tonne of beryl flour. Thus with a 25 mass percent excess, that correspond to a practical sulfuric acid-to-solid mass ratio [H_2SO_4-to-S] of 1,358 kg (100 wt.% H_2SO_4) per tonne of beryl flour.

The key figures related to the sulfation of beryl are summarized and reported in Table 52 hereafter.

Table 52 – Sulfation of beryl: key figures

BERYL (concentrate)			
Chemical composition		Strength H_2SO_4 (initial)	**77.7%**
SiO_2	67.00 wt.%	Strength H_2SO_4 (attack)	77.7%
Al_2O_3	18.90 wt.%	[H_2SO_4-to-Solid](Theor.)	**1,087** kg/tonne
BeO	13.80 wt.%	Mass excess of acid	**25%**
	99.70 wt.%	[H_2SO_4-to-Solid](Actual)	**1,358** kg/tonne
		[Acid-to-Solid](Actual)	**1,748** kg/tonne
		Solids pulp density	36%
		c_p (solid)	795 J/kg/K
		c_p (acid)	1643 J/kg/K
		c_p (pulp)	**1179** J/kg/K
		$\Delta h_{Sulfation}$	**-448** kJ/kg of solid
Mass of solids (dry)	**1000** kg	$\Delta h_{Hydration}$	0 kJ/kg of acid
Mass of sulfuric acid	1748 kg	$\Delta h_{Vaporization}$	2256 kJ/kg of steam
Mass of water (injected)	0 kg	Δh_{Total}	**157** kJ/kg of mixture
Mass of pulp	**2748** kg	Adiabatic temp rise ΔT	**-133 K ENDOTHERMIC**

Because the sulfation reaction of the beryl flour is endothermic despite the beryllia content (up to 11.6 wt.% BeO) thus the mixture needs to be heated above 120°C inside a kiln to trigger the sulfation reaction. The stoichiometric chemical reaction is described below:

$$Be_3Al_2(Si_2O_6) + 6H_2SO_4(l) = 3BeSO_4 + Al_2(SO_4)_3 + 2SiO_2(s) + 3H_2O(g)$$

Once the sulfation reaction reach the operating temperature range between 200°C and 250°C, the sulfation baking is maintained during for 3 to 4 hours. After cooling down to 80°C the sulfation cake is dissolved in warm water. The silica residue is removed by filtration. Then ammonium sulfate is added to the clear pregnant solution in order to crystallize the aluminum sulfate removed by centrifugation.

Afterwards, the beryllium-rich pregnant liquor is concentrated by evaporation and the beryllium sulfate is crystallized out by subsequent cooling. After calcination of the beryllium sulfate at 1400°C for 2 hours, pure beryllium oxide (beryllia), BeO, CAS No. [1304-56-9], is obtained and can be used as intermediate for the production of beryllium metal by molten salt electrolysis.

9.9 Rare Earths and Thorium from Monazite

The sulfuric acid digestion of monazite also called sulfation baking was done for more than 100 years. **Monazite** is an ubiquitous cerium-rich phosphate mineral often hosting significant concentration of thorium with the empirical chemical formula $(Ce,Pr,Nd,Th)PO_4$ [89] and which is recovered industrially as a by-product during the beneficiation of beach sand ilmenite (mineral sands) in Florida, South Africa, Madagascar, Travancore, and Orissa States India, and Western Australia.

Historically, the sulfation baking process of monazite has been first used in France by *La Société des produits chimiques des terres rares* created in 1919 by Professor Georges Urbain associated with Joseph Blumenfeld, and Félix Binder, who bought the assets of *Deutsche Auergesellschaft*. The plant was first located in Serquigny (Eure) for producing cerium and thorium salts for the manufacture of Auer gas mantles [90]. Following its total destruction during WWII, a new plant was built near La Rochelle in 1948 and then owned by *Rhône-Poulenc* later renamed *Rhodia* and today owned by the Belgium group *Solvay* [91].

It is important to mention that the sulfuric acid digestion of monazite was always in competition with the caustic digestion or caustic cracking, the latter method being used for processing higher grade of monazite containing at least 70 wt.% Ln_2O_3, where Ln denotes ceric lanthanides, due to the significant higher operating cost of using caustic soda instead of concentrated sulfuric acid.

The sulfation baking is currently used by *Lynas Corporation* at its Malaysian plant in Gebeng, while the processing has been performed in the United State during the 1970, in Australia, in Brazil and India. For the latter two countries, the processing of monazite was performed essentially to recover the thorium values to

[89] ANTHONY, J.W.; BIDEAUX, R.A.; BLADH, K.W.; and NICHOLS, M.C. (2000) *Handbook of Mineralogy, Volume IV: Arsenates, Phosphates, Vanadates.* Mineral Data Publishing, Tucson, AZ, pp. 382-384.

[90] STOSKOPF, N. (2005) Trois aspects d'une crise industrielle: La Fabrique de produits chimiques Thann & Mulhouse (1974-1983). Colloque de Metz, Metz, France. pp. 119-131.

[91] ROLLAT, A. (2016) Recovery of rare earths from wet-process phosphoric acid, the Solvay experience. *Procedia Engineering,* **138**, 273-280.

support both Brazil, and India nuclear programs thus forbidding the export of thorium and thus favoring the export the rare earths instead.

In it is interesting to note that recently several researchers have revisited this ancillary process [92, 93, 94] investigating the impact of various operating conditions such as the sulfuric acid concentration, the sulfation baking temperature, and the digestion duration including the recent review article from Demol et al. describing and comparing the processes for sulfuric acid baking of monazite, xenotime, and other rare earth minerals [95].

Regarding the numerous junior mining projects that are currently under development globally for the possible production of rare earth from various ores and techniques including sulfation techniques discussed here, we suggest to read the review articles from Verbaan et al. [96].

A schematic flow diagram depicting the old (historical) sulfuric acid cracking of monazite is depicted in Figure 24.

Industrially, during the sulfation baking, the monazite concentrate is first ground in ball mills to less than 65 mesh Tyler (e.g., 212 μm), then the ground concentrate is digested with 93 wt.% sulfuric acid inside a stirred reactor at 210°C for 4 hours with a sulfuric acid-to-solid mass ratio [H_2SO_4-to-S] of nearly 1,560 kg (100 wt.% H_2SO_4) per tonne of monazite. The operation was conducted historically batch wise inside cast iron pots, later in stirred brick-lined reactors.

More recently the sulfation roasting is performed inside long refractory-lined rotary kilns such as the four 60-meter long kiln currently used by *Lynas Corp*. Actually, *Lynas Advanced Material Plant* (LAMP) is the world's largest facility and rare-earth processing factory located in Gebeng, Pahang, Malaysia. This facility is designed to treat monazite concentrate mined from Mount Weld, Western Australia. The monazite concentrate is then transported and shipped to the company chemical facility in Gebeng, Malaysia to undergo first the sulfuric acid cracking and then the rare earth separation process.

---

[92] DEMOL, J.E.; HO, E.; and SENANAYAKE, G. (2018) Sulfuric acid baking and leaching of rare earth elements, thorium and phosphate from a monazite concentrate: Effect of bake temperature from 200 to 800°C. *Hydrometallurgy*, **179**(8)254-267.

[93] BERRY, L.; AGARWAL, V.; GALVIN, J.; and SADEGH SAFARZADEH, M. (2018) Decomposition of monazite concentrate in sulphuric acid. Canadian Metallurgical Quarterly, **57**(4)422-433

[94] SADRI, F.; RASCHI, F.; and AMINI, A. (2017). Hydrometallurgical digestion and leaching of Iranian monazite concentrate containing rare earth elements Th, Ce, La and Nd. *International Journal of Mineral Processing*, **159**, 7-15.

[95]J DEMOL, J.; HO, E.; SOLDENHOFF, K.; and SENANAYAKE, G. (2019) The sulfuric acid bake and leach route for processing of rare earth ores and concentrates: A review. *Hydrometallurgy*, **188**(9)123-139.

[96] VERBAAN, N.; BRADLEY, K.; BROWN, J.; and MACKIE, S. (2015) A review of hydrometallurgical flowsheets considered in current REE projects. In: SIMANDL, G.J. and NEETZ, M., (Eds.), *Symposium on Strategic and Critical Materials Proceedings*, November 13-14, 2015, Victoria, BC. British Columbia Ministry of Energy and Mines, British Columbia Geological Survey Paper 2015-3, pp. 147-162.

A schematic flow diagram depicting the modern sulfuric acid cracking of monazite is depicted in Figure 25.

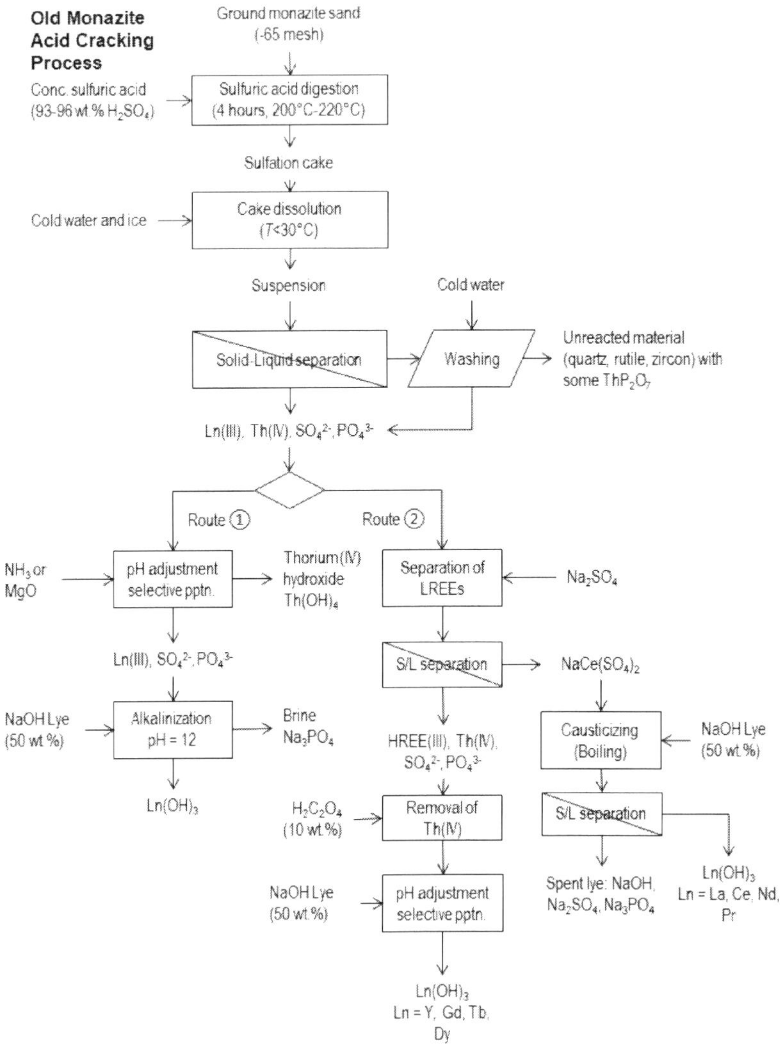

Figure 24 – Flow diagram for the sulfuric acid-cracking of monazite (Old process)

During the sulfation process, it is important that the temperature should be kept always below 230°C in order to prevent the formation of water-insoluble thorium pyrophosphate with chemical formula ThP_2O_7.

Due to the high temperature maintained during roasting some unescapable evolution of sulfur dioxide together with steam occurs and the latter must be scrubbed after entrainment dust has been removed with Cottrell's electrostatic precipitators.

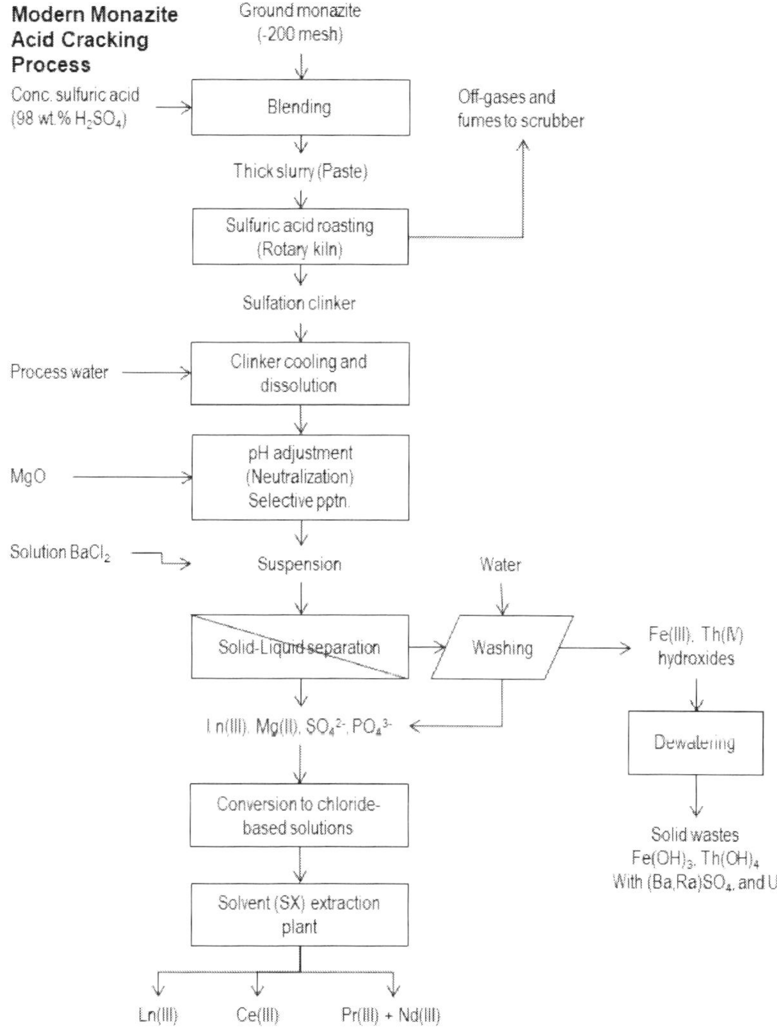

Figure 25 – Flow diagram for the sulfuric acid-cracking of monazite (Modern process)

For instance, the sulfation reaction for a monazite exhibiting an averaged chemical composition of 27 wt.% P_2O_5, 60 wt.% Ln_2O_3 (including cerium expressed as Ce_2O_3) with Ln referring to a lanthanide element (*lanthanon*) with average molar mass of 140 kg/kmol and 13 wt.% ThO_2 that corresponds to the empirical chemical formula $(Ln_{0.96}Th_{0.13})PO_{4.2}$ proceeds according to the following reaction scheme:

$$Ln_{0.96}Th_{0.13}PO_{4.20} + 1.70H_2SO_4 = 0.48Ln_2(SO_4)_3 + 0.13Th(SO_4)_2 + H_3PO_4 + 0.20H_2O$$

The above chemical reaction shows that the theoretical sulfuric acid-to-solid mass ratio [H_2SO_4-to-S] is 619 kg (100 wt.% H_2SO_4) per tonne of dry monazite concentrate in practice based on the variation in chemical compostion it ranges between 620 and 640 kg (100 wt.% H_2SO_4) per tonne.

Because the sulfation reaction is strongly exothermic (-680 kJ/kg) in order to keep the maximum operating temperature around 210°C at all time while maintaining optimum rheological conditions of the slurry with a pasty consistency, it is necessary to use an excess of sulfuric acid ranging from 200 to 230 mass percent.

Therefore, the practical (actual) sulfuric acid-to-solid mass ratio [H_2SO_4-to-S] ranges between 1,600 and 2,000 kg (100 wt.% H_2SO_4) per tonne of monazite sand. These operating conditions ensure that the completion of the sulfation reaction reaches at least 95%.

The key figures related to the sulfation of monazite according to the modern sulfation process are summarized and reported in Table 53 hereafter.

Table 53 – Sulfation of monazite: key figures

MONAZITE (concentrate)			
Chemical composition		Strength H_2SO_4 (initial)	**98%**
La_2O_3	62.70 wt.%	Strength H_2SO_4 (attack)	98%
P_2O_5	28.00 wt.%	[H_2SO_4-to-Solid](Theor.)	**635** kg/tonne
ThO_2	4.52 wt.%	Mass excess of acid	**200%**
SiO_2	2.30 wt.%	[H_2SO_4-to-Solid](Actual)	**1,906** kg/tonne
CaO	0.90 wt.%	[Acid-to-Solid](Actual)	**1,945** kg/tonne
Fe_2O_3	0.45 wt.%	Solids pulp density	34%
Al_2O_3	0.40 wt.%	c_p (solid)	649 J/kg/K
	99.27 wt.%	c_p (acid)	1418 J/kg/K
		c_p (pulp)	**1144** J/kg/K
		$\Delta h_{Sulfation}$	**-680** kJ/kg of solid
Mass of solids (dry)	**1000** kg	$\Delta h_{Hydration}$	-63 kJ/kg of acid
Mass of sulfuric acid	1945 kg	$\Delta h_{Vaporization}$	2256 kJ/kg of steam
Mass of water (injected)	0 kg	Δh_{Total}	**-242** kJ/kg of mixture
Mass of pulp	**2945** kg	Adiabatic temp rise ΔT	**212 K EXOTHERMIC**

During the sulfation baking process most of the thorium, rare earths, and uranium are fully converted to the corresponding metal sulfates. After solidification upon cooling down to 70°C, the thick paste obtained is water soluble and it is dissolved in a cold mixture of ice and water.

The practical water-to-paste mass ratio [H_2O-to-P] ranges from 2,200 to 2,900 kg of water per tonne of sulfated mixture, that is, a water-to-monazite mass ratio ranging 8,000 kg to 10,000 kg per tonne of monazite sand. The freezing condition above is mandatory owing to the higher solubilities of lanthanide sulfates at low temperature [97].

Actually not only lanthanide sulfate hydrates exhibits a minimum solubility near the middle of the series but they also have their solubility reduced by half when the temperature is increased from 20°C to 40°C as reported in Table 54 [98].

[97] FLAHAUT, J. (1969) *Les elements des terres rares*. Collection de Monographies de Chimie, Masson & Cie Éditeurs, Paris, France, pp. 38-40.

[98] YOST, D.M., RUSSEL, H. Jr.; and GARNER, C.S. (1947) *The Rare Earth Elements & Their Compounds*. Wiley, New York, NY, pp 42.

Table 54 – Solubility of lanthanide sulfate hydrates in water

	Solubility in grams per 100 grams of water	
Temperature	20°C	40°C
$La_2(SO_4)_3.9H_2O$	2.20	1.18
$Ce_2(SO_4)_3.8H_2O$	9.43	5.70
$Pr_2(SO_4)_3.8H_2O$	12.74	7.64
$Nd_2(SO_4)_3.8H_2O$	7.00	4.51
$Sm_2(SO_4)_3.8H_2O$	2.67	1.99
$Eu_2(SO_4)_3.8H_2O$	2.56	1.93
$Gd_2(SO_4)_3.8H_2O$	2.89	2.19
$Tb_2(SO_4)_3.8H_2O$	3.56	2.51
$Dy_2(SO_4)_3.8H_2O$	5.07	3.34
$Ho_2(SO_4)_3.8H_2O$	8.18	4.52
$Y_2(SO_4)_3.8H_2O$	9.76	4.90
$Er_2(SO_4)_3.8H_2O$	16.00	6.53
$Yb_2(SO_4)_3.8H_2O$	34.78	22.90
$Lu_2(SO_4)_3.8H_2O$	47.27	16.93

This yields after settling, a pregnant leach solution while insoluble silica, zircon, and rutile together with unreacted monazite form a dense sludge at the bottom of the reactor. The denser unreacted monazite is separated by filtration or centrifugation from the sludge and eventually recycled upstream or disposed-off.

Radium which is always present as the 226-Ra radionuclide is a deleterious and extremely radioactive impurity due to its short half-life (ca. 1,600 years [99]), thus it is usually removed from the clear pregnant solution by simply adding incrementally either a saturated aqueous solution of barium chloride ($BaCl_2$) or barium carbonate ($BaCO_3$).

The insoluble barium sulfate ($BaSO_4$) that precipitates readily removes all the radium cations by co-precipitation as insoluble intimate mixture of barium and radium sulfates that are separated by filtration or gravity settling.

$$(1 - x)Ba^{2+} + x^{226}Ra^{2+}(traces) + SO_4^{2-} = (Ba_{1-x}^{226}Ra_x)SO_4(s)$$

This process produces a clean solution of thorium, rare-earths, and uranium cations together with both sulfate and phosphate anions. Because of the

---

[99] LAGOUTINE, F.; and COURSOL, N. (1986) Période radioactives. table des valeurs recommandées. Bureau National de Métrologie (BNM), Editions Chiron, Paris, France.

chemical similarity existing between lanthanides and thorium cations in solution, it is difficult to separate thorium from lanthanides in the presence of phosphate anions [100].

The successful separation of thorium from lanthanides is tedious and thus several different routes have been devised to address this peculiar problem. Each route can be used whether the recovery of thorium or the recovery of lanthanide is targeted [101].

When the recovery of thorium is the main target and the recovery of lanthanide only optional, the addition of concentrated hydrofluoric acid (HF) to the pregnant solution is performed in order to precipitate all the thorium values as thorium fluoride (ThF$_4$). Despite the well-known low solubility of lanthanides fluorides (LnF$_3$) in aqueous solutions, such successful separation can be achieved very efficiently only because of the strong acidity prevailing due to the high concentration of phosphoric acid and excess sulfuric acid left behind that preclude their co-precipitation.

Conversely, if the recovery of lanthanides is the goal, in this particular case, the selective precipitation of thorium hydroxide [Th(OH)$_4$] is conducted first by simply raising the pH to 1.0 either by dilution with water or by adding a neutralizing agent. The latter is achieved by sparging gaseous ammonia (NH$_3$) directly into the pregnant leach solution.

After removal of the thorium hydroxide by filtration, the precipitation of the lanthanide hydroxides is obtained by raising the pH to 2.5 by either adding soda ash (i.e., sodium carbonate) or caustic soda (i.e., sodium hydroxide) or as oxalates by adding oxalic acid.

The next step is the selective recovery of rare earths. It was performed historically by selective precipitation [102] or since the 1950s by solvent extraction (SX) from concentrated nitric solutions of Th(IV) using tri-butyl phosphate (TBP) as extracting agent, the latter dissolved in kerosene to improve the difference in densities between the organic (light) and aqueous (heavy) phases for a fast demixing and reduced frothing [103].

[100] WYLIE, A.W. (1959) Extraction and purification of thorium. *Reviews in Pure and Applied Chemistry*. **9**(3)169-212.

[101] FLAHAUT, J. (1969) *Les éléments de terres rares*. Collection de monographie de chimie, Masson & Cie, Paris, France.

[102] STOLL, W. (1997) Chapter 42. Thorium - From Habashi, F. (ed) (1997) *Handbook of Extractive Metallurgy*. Wiley VCH, Weinheim, Germany, pp. 1656-1660.

[103] BENEDICT, M.; PIGFORD, T.H.; and LEVY, H.W. (1981) *Nuclear Chemical Engineering, Second Edition*. McGraw-Hill Book Company, New York, NY, pp. 307-309.

9.10 Niobium from Pyrochlore

The most important niobium minerals of commercial importance mined today belong to the pyrochlore, and in a lesser extent, minerals from the columbite group.

The ***pyrochlore group minerals*** [104, 105] exhibit a cubic crystal lattice structure having the general formula: $AB_2O_6(O,OH,F)$ where $A = K^+$, Cs^+, Ca^{2+}, Sr^{2+}, Ba^{2+}, Ln^{3+}, Pb^{2+}, and U^{4+}; and $B = Nb^{5+}$, Ta^{5+}, Ti^{4+}, Sn^{4+} and W^{6+}. The ***pyrochlore*** (*sensu stricto*) $[(Na,Ca)_2(Nb,Ta,Ti)_2O_4(OH,F).H_2O$ is of paramount commercial importance as it is by far the main niobium ore.

The ***columbite group minerals*** [106] exhibit an orthorhombic crystal lattice structure having the general formula AB_2O_6 wherein $A = Fe^{2+}$, Mn^{2+} and Mg^{2+}; and $B = Nb^{5+}$ and Ta^{5+} with complete isomorphic substitutions between the end members. The niobium-rich end is named ***ferrocolumbite*** $[(Fe,Mn)Nb_2O_6]$.

The two major niobium producers worldwide are *Companhia Brasileira de Metalurgia e Mineração* (CBMM) at Araxa (M.G.), Brazil with 90 percent of the world market and *Niobec* (now *Iamgold*) in Saint-Honoré de Chicoutimi, Qc, Canada, supplying the remaining 10 percent. Both companies process the pyrochlore concentrate using an aluminothermic process to produce ferroniobium.

Nowadays, the production of niobium chemicals is obtained by the so-called HF-process such as the one performed by *H.C. Starck GmbH* at Goslar in Germany and formerly *Cabot Performance Materials* in Boyerton, USA. The tantalo-columbite concentrate is first subjected to a chemical dissolution step using mixtures of concentrated hydrofluoric and sulfuric acids (HF-H_2SO_4). The decomposition step is followed by a solid-liquid separation step order to remove the insoluble residues. The clear pregnant leach solution is used for performing the separation of tantalum and niobium by solvent extraction (SX).

However, in the past pyrochlore concentrates have been also processed commercially by two companies using various sulfation techniques. These two sulfation processes are described hereafter.

[104] HOGARTH, D.D. (1977) Classification and nomenclature of the pyrochlore group. *American Mineralogist*, **62**, 403-410.

[105] ANTECIO, D. (2021) Pyrochlore-supergroup minerals nomenclature: An update. *Frontier in Chemistry*, **9**, 713368

[106] ERCIT, T.S.; WISE, M.A.; and CERNY, P. (1995) Compositional and structural systematics of the columbite group. *American Mineralogist*, **80**, 613-619.

9.10.1   The IG-Farbenindustrie Process

The sulfation process was performed commercially at the *IG Farbenindustrie AG* plant in the 1950s located in Bitterfeld (Wolfen) back then East Germany [107]. The pyrochlore ore containing only 0.5 wt.% Nb_2O_5 was first leached with nitric acid (HNO_3) to remove most of the calcium as calcium nitrate, $Ca(NO_3)_2$, used in the fertilizer industry while leaving an insoluble solid residue containing only 5 wt.% Nb_2O_5. This step was mandatory as otherwise, the sulfation of the raw ore would have produced on-site large amount of gypsum of no commercial value.

The insoluble residue was then further beneficiated by gravimetric and magnetic methods to produce a higher grade concentrate containing 15-17 wt.% Nb_2O_5. The pyrochlore concentrate was then digested with 60°Be (78 wt.% H_2SO_4) concentrated sulfuric acid produced from the Glover Tower for 4 hours at 180°C. The reported practical (actual) sulfuric acid-to-niobium pentoxide mass ratio [H_2SO_4-to-Nb_2O_5] was 40 kg of sulfuric acid (100 wt.% H_2SO_4) per kg of Nb_2O_5 that corresponds to an elevate liquid-to-solid mass ratio [A-to-S] of nearly 27,815 kg of sulfuric acid (78 wt.% H_2SO_4) per tonne of pyrochlore concentrate.

The key figures related to the sulfation of pyrochlore according to the IG-Farnen process are summarized and reported in Table 55 hereafter.

---

[107] SAMSONOV, G.V.; and KONSTANTINOV, V.I. (1959) *Tantal I Niobiy*. Literatury po Chernoy I Tsvernoy Metallurgii. Gosudarstvennoye Nauchno-Tekhnicheskoye Izdatel'STVO, Moscow, Russia.

Table 55 – Sulfation of pyrochlore (IG-Farben): key figures

PYROCHLORE (conc.) [IG Farben]			
Chemical composition		Strength H_2SO_4 (initial)	**78%**
Nb_2O_5	53.00 wt.%	Strength H_2SO_4 (attack)	78%
CaO	16.00 wt.%	[H_2SO_4-to-Solid](Theor.)	**834** kg/tonne
FeO	4.00 wt.%	Mass excess of acid	**2500%**
SiO_2	3.00 wt.%	[H_2SO_4-to-Solid](Actual)	**21,696** kg/tonne
TiO_2	5.00 wt.%	[H_2SO_4-to-Nb_2O_5](Actual)	**40,935** kg/tonne
Na_2O	3.00 wt.%	[Acid-to-Solid](Actual)	**27,815** kg/tonne
	84.00 wt.%	Solids pulp density	3%
		c_p (solid)	320 J/kg/K
		c_p (acid)	1639 J/kg/K
		c_p (pulp)	**1362** J/kg/K
		Δh Sulfation	**-1202** kJ/kg of solid
Mass of solids (dry)	**1000** kg	Δh Hydration	0 kJ/kg of acid
Mass of sulfuric acid	27815 kg	Δh Vaporization	2256 kJ/kg of steam
Mass of water (injected)	0 kg	Δh Total	**437** kJ/kg of mixture
Mass of pulp	**28815** kg	Adiabatic temp rise ΔT	**-321 K ENDOTHERMIC**

The liquor was then diluted with large volumes of water to bring the final acid concentration down to 55 wt.% H_2SO_4. The unreacted solid residue was then removed by filtration and the clear pregnant solution was diluted again down to 20 wt.% H_2SO_4 and neutralized with gas ammonia (NH_3) to precipitate the niobium hydroxide, $Nb(OH)_5$ while maintaining a free sulfuric acid concentration of at least 2 wt.%. This residual acidity was suitable to prevent the co-precipitation of Ti(IV), and Fe(III). The niobium hydrolysate precipitate was filtered, thoroughly washed, and oven-dried. After calcination, the final product contained 96 wt.% Nb_2O_5 with 2 wt.% Fe_2O_3, and 0.5 wt.% TiO_2. The final product was further refined by using the HF-process by *H.C. Starck GmbH* at its plant in Goslar, Germany.

9.10.2 Société Française d'Électrométallurgie

The French chemist and Professor at the *École Supérieure de Physique et de Chimie Industrielle* (ESPCI) Gaston Charlot was the first to investigate and invent in the early 1970s the direct sulfation of pyrochlore concentrates. The process was patented in 1976 by the *Société Française d'Électrométallurgie* (SOFREM) [108]. A schematic of the process flow diagram is depicted in Figure 26.

---

[108] CHARLOT, G. (1976) *Process for treatment of pyrochlore concentrates.* U.S. Patent 3,947,542 (Société Française d'Électrométallurgie), March, 30, 1976.

SOFREM Process

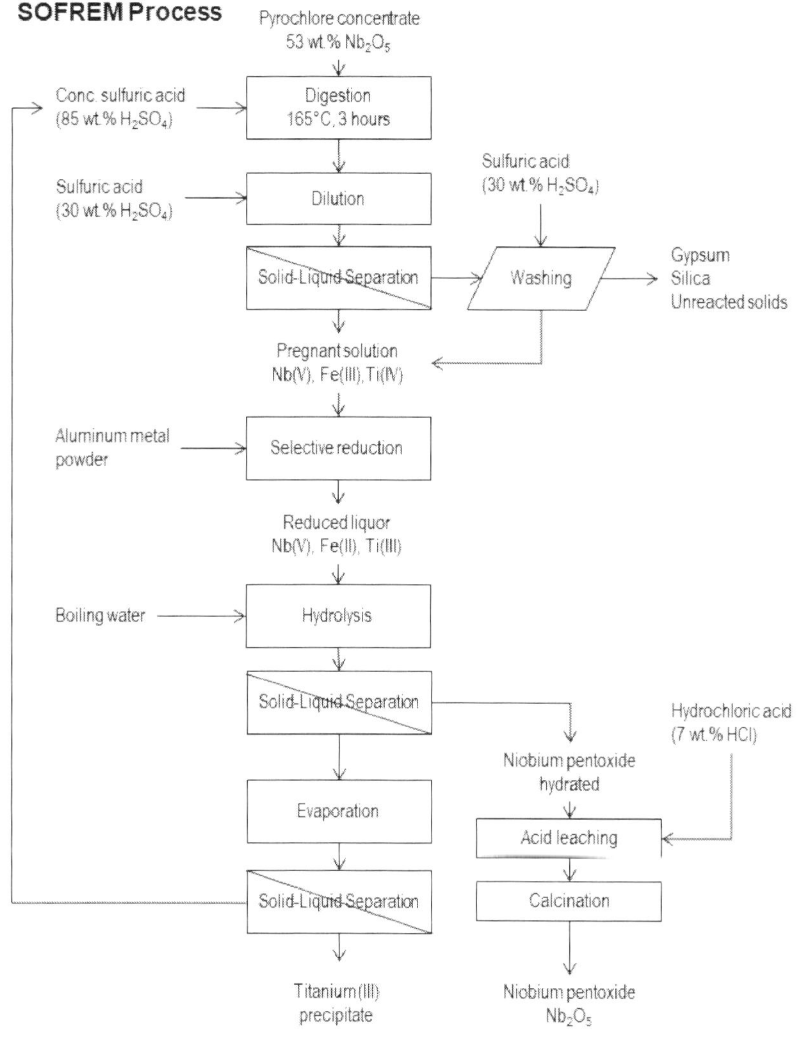

Figure 26 – Flow diagram for the Sofrem process

The process was tested at the pilot scale on batches of 1,000 kg of pyrochlore concentrate exhibiting the following average chemical composition listed in Table 56.

Table 56 – Chemical composition of pyrochlore concentrate

Oxide	Mass percentage (wt.%)
Nb_2O_5	53
CaO	16
TiO_2	5
FeO	4
SiO_2	3
Na_2O	3

The first step consisted to perform the sulfuric acid digestion of the finely ground pyrochlore concentrate with concentrated sulfuric acid (85 wt.%H_2SO_4) at 165°C. The stoichiometric (theoretical) sulfuric acid-to-solid mass ratio [H_2SO_4-to-S] is only 834 kg (100 wt.% H_2SO_4) per tonne of dry pyrochlore concentrate.

Because the extremely high enthalpy of the sulfation reaction (-1200 kJ/kg) mostly related to the elevate calcium oxide (calcia) content, 18 times the theoretical amount of 100 wt.% sulfuric acid were used that corresponds to the reported sulfuric acid-to-niobium pentoxide mass ratio [H_2SO_4-to-Nb_2O_5] of almost 30 kg of sulfuric acid (100 wt.% H_2SO_4) per kg of Nb_2O_5. This extreme dilution yielded a a solid pulp density of only 3 mass percent.

Therefore, the practical (actual) liquid-to-solid mass ratio [A-to-S] used was close to 18,000 kg of sulfuric acid (85 wt.%H_2SO_4) per kg of pyrochlore concentrate. After cooling down to 40°C, the slurry was diluted further with 30 wt.% sulfuric acid recycled from the downstream process. The weaker sulfuric acid-to-slurry mass ratio was 1,000 kg of sulfuric acid per 3 tonnes of slurry.

The key figures related to the sulfation of pyrochlore according to the SOFREM process are summarized and reported in Table 55 hereafter.

Table 57 – Sulfation of pyrochlore (SOFREM): key figures

PYROCHLORE (conc.) [SOFREM]			
Chemical composition		Strength H_2SO_4 (initial)	**85%**
Nb_2O_5	53.00 wt.%	Strength H_2SO_4 (attack)	85%
CaO	16.00 wt.%	[H_2SO_4-to-Solid](Theor.)	**834** kg/tonne
FeO	4.00 wt.%	Mass excess of acid	**2000%**
SiO_2	3.00 wt.%	[H_2SO_4-to-Solid](Actual)	**17,523** kg/tonne
TiO_2	5.00 wt.%	[H_2SO_4-to-Nb_2O_5](Actual)	**33,063** kg/tonne
Na_2O	3.00 wt.%	[Acid-to-Solid](Actual)	**20,616** kg/tonne
	84.00 wt.%	Solids pulp density	5%
		c_p (solid)	320 J/kg/K
		c_p (acid)	1554 J/kg/K
		c_p (pulp)	**1349** J/kg/K
		$\Delta h_{Sulfation}$	**-1202** kJ/kg of solid
Mass of solids (dry)	**1000** kg	$\Delta h_{Hydration}$	0 kJ/kg of acid
Mass of sulfuric acid	20616 kg	$\Delta h_{Vaporization}$	2256 kJ/kg of steam
Mass of water (injected)	0 kg	Δh_{Total}	**267** kJ/kg of mixture
Mass of pulp	**21616** kg	Adiabatic temp rise ΔT	**-198 K ENDOTHERMIC**

The unreacted solids, gypsum, and precipitated metal sulfates were separated by filtration and washed with fresh 72 wt.% sulfuric acid to recover all the niobium values. The pregnant solution was then reduced by adding an excess of coarse (1-2 mm) aluminum metal powder in order to reduce all the Ti(IV) and Fe(III) into Ti(III) and Fe(II) respectively thus preventing the hydrolysis and co-precipitation of Ti(IV) with Nb(V). The reduction was performed during three hours under inert nitrogen atmosphere to prevent air oxidation.

The reduced solution was then mixed with boiling water to trigger the precipitation of niobium hydrate that was recovered by filtration, washed with 2M hydrochloric acid to remove traces of lanthanides and thorium, thoroughly washed with hot deionized water, oven-dried, and finally calcined to yield pure niobium pentoxide (99.5 wt.% Nb_2O_5). The niobium pentoxide recovery rate achieved in that way was 95.8%.

Afterwards, the filtered barren solution was combined with the wash waters, and concentrated by thermal evaporation until the sulfuric acid concentration reached 60 wt.%. H_2SO_4. A titanium (III) precipitate was then obtained and recovered. Further evaporation was performed to produce a 85 wt.% acid to be reused upstream.

9.11 Sodium and Potassium Sulfates by the Mannheim Process

Nowadays hydrogen chloride (HCl) and hydrochloric acid (i.e., aqueous solutions of HCl) are mainly obtained industrially by the direct combustion of pure hydrogen gas with chlorine gas inside a carbon-lined reactor. For example, such complete units for the direct synthesis hydrochloric acid are built by the German company *SGL Carbon GmbH*. Both gases being produced by the chlor-alkali process while significant amount of HCl is also by-produced from the chlorination of olefins such as the production of polyvinyl chloride (PVC).

However, historically the major source of HCl was produced by the so-called ***Manheim process*** that consisted to react sodium chloride (NaCl), also called ***rock salt*** in the trade, with concentrated sulfuric acid [109, 110]. The Manheim process was also the first mandatory stage in the ***LeBlanc process*** where the anhydrous sodium sulfate obtained was afterwards roasted with carbon and limestone to yield soda ash.

The two chemical reactions occurring during the process are divided in two consecutive steps.

The first step consists to the formation of ***sodium hydrogenosulfate*** (i.e., sodium bisulfate) according to the following chemical reaction which is highly exothermic reaching a maximum temperature of 150°C:

$$NaCl(s) + H_2SO_4(l) = NaHSO_4(s) + HCl(g)$$

Then, additional sodium chloride is added to the molten sodium bisulfate and roasted at a much higher temperature of 600°C or higher reacting according to the second chemical reaction as follows:

$$NaCl(s) + NaHSO_4(l) = Na_2SO_4(s) + HCl(g)$$

The Mannheim process was performed inside a staged chambers reactor called the *Maletra furnace* [111].

First the rock salt and concentrated sulfuric acid (60°Be) produced from the Glover tower were mixed slowly together inside moderately heated lead-lined

[109] von PLESSEN, H. (2000) *Sodium Sulfates*. In *Ullmann's Encyclopedia of Industrial Chemistry*. Wiley-VCH, Weinheim, Germany.

[110] BUTTS, D.; and BUSH, D.R. (2000) *Sodium Sulfates and Sulfides*. In Kirk-Othmer Encyclopedia of Chemical Technology, John Wiley & Sons, Inc., New York, NY.

[111] MOLINARI, E. (1943) *Trattato di chimica inorganica generale ed applicata all'industria, sesta edizione, Volume I Tomo I e II*. Ulrico Hoepli Editore, Milano, Italy.

pots or most often cast iron flat pans at a temperature below 150°C by means of hot gases from the hearth.

After a certain residence time, all the first HCl gas that has evolved exited the top leaving behind a pasty solid mass called the **salt cake** made of anhydrous sodium bisulfate. The salt cake was moved downward by rakes onto a lower pan.

The lower portion of the furnace was brick-lined with acid-resistant refractories and gas flame heated. The inside temperature was maintained above 600°C to ensure that all the anhydrous sodium bisulfate (*m.p.* 315°C) was in the molten state exhibiting a high fluidity suitable to allow efficient mixing between reactants. The remaining portion of sodium chloride was then added to complete the reaction and to yield solid sodium sulfate (*m.p.* 890°C) as described above.

The hot flue gases existing from the top consisted mainly of hydrogen chloride gas. After the entrained particulates were removed inside baffled boxes, the hot gases were cooled-down, scrubbed with concentrated sulfuric acid, and then absorbed into water inside a packed tower filled with Vycor® packing internals producing concentrated hydrochloric acid (32 wt.% HCl).

On the other hand, the anhydrous sodium sulfate was discharged from the bottom, cooled down onto steel drums, and the lumps milled, screened, and bagged ready to be used in the detergent and glass making industry.

The Maletra furnace consisted of staged chambers where the reactants flow from top to bottom while HCl gas escape from the top.

The Mannheim process was historically devised for the production of sodium sulfate, but nowadays, the Mannheim process is still used especially in Europe by *Tessenderlo Group* (Belgium) and several private producers in mainland China to produce **potassium sulfate** (K_2SO_4) from potassium chloride (KCl) also called the **muriate of potash** (SOP) in the fertilizer industry. It is estimated that 50 percent of the global supply of potassium sulfate is produced by such process

However, this process due to high energy consumption has the highest operating cost in the industry ranging from \$400 to \$550 per tonne of potassium sulfate produced when compared to other traditional production routes (e.g., crystallization from complex brines, mixed salts reaction).

9.12 Epsom Salt from Magnesia and Magnesite

Magnesium sulfate heptahydrate, $MgSO_4.7H_2O$, also called **Epsom salt** in the trade can be produced by the direct sulfation of low iron dolime (i.e., calcine dolomite: MgO+CaO), upon completion of the strongly exothermic reaction, the soluble magnesium sulfate is leached with hot slightly acidic water leaving the insoluble gypsum and/or anhydrite the latter depending on the initial sulfuric acid strength and operating temperature. The process was invented by Hendel and performed in the USA by the *Wigton-Abbot Corp.* [112] using spent sulfuric acid and waste oleum both contaminated with deleterious organic pollutants originating from the manufacture of DDT.

After cross flow filtration, the clear pregnant leach solution containing between 350-400 g/L of $MgSO_4$ is concentrated by multi-stage vacuum evaporation, and the crystallized Epsom salt is recovered by centrifugation, followed by mild drying to obtain dry crystals.

In the past 50 years, several attempts were to perform the sulfation of other magnesium-rich feedstock such as olivine, peridotite, serpentine or even asbestos tailings. All these routes have been explored all with aim to produce concentrated brine from which the magnesium salt is obtained as above.

9.13 The PHAR Process

Green Technology Group (Sharon, CT) in collaboration with *Chester Engineers* (Pittsburgh, PA), and *US Steel Group* (Davosburg, PA), started a project in February 2000 in the United States, with some financial support from the NICE[3] program under the auspices of the *U.S. Department of Energy* (DOE) and the *U.S. Environmental Protection Agency*(EPA) with the aims to demonstrate and commercialize a new process called **Pickliq Hydrochloric Acid Regeneration (PHAR)** [113] devoted to the regeneration of hydrochloric acid from ferrous chloride contained in spent pickling liquors (SPLs).

---

[112] HENDEL, F.J. (1956) *Process for treating waste acid.* U.S. Patent 2,754,175 (Wigton-Abbot Corporation), July 10th, 1956.

[113] COLLECTIVE (2000) *Energy-saving Regeneration of Hydrochloric Acid Pickling Liquor.* U.S. Department of Energy (DOE), Energy Efficiency and Renewable Energy, June 2000.

**Pickliq Hydrochloric Acid
Regeneration (PHAR) Process**

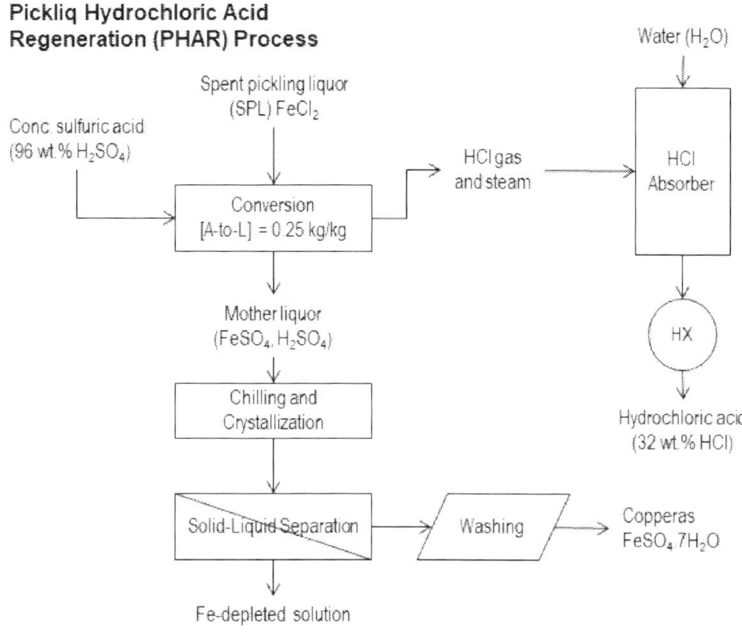

Figure 27 – Flow diagram for the PHAR recycling process

Basically, the PHAR process allows the recycling of both tainted hydrochloric acid and spent pickling liquors. After a filtration step, in order to remove insoluble residues, the spent pickling liquor is pumped into a stirred chemical reactor where concentrated sulfuric acid (93-98 wt.% H_2SO_4) is added in order to react with ferrous chloride to produce ferrous sulfate and evolves hydrogen chloride gas according to the simple chemical reaction:

$$FeCl_2 + H_2SO_4 = FeSO_4 + 2HCl$$

The concentrated sulfuric acid-to-spent pickling liquor mass ratios ranges usually between 0.33 kg per kg to 0.20 kg per kg. Because of the highly exothermic reaction of hydration of sulfuric acid, most the hydrogen chloride (HCl) is evolved as gas mixed with steam. The vapor mixture is feed directly to a countercurrent absorption tower (absorber) for producing fresh concentrated hydrochloric acid (32 wt.% HCl) which is sent back to the pickling process. Afterwards, the remaining iron-rich solution is chilled between -12°C and 0°C, causing the crystallization of ferrous sulfate heptahydrate.

The aqueous suspension of copperas crystals or slurry is pumped by a diaphragm pump and the crystals of salt are removed from the solution by centrifugation, washed with acidic cold water and dried. The remaining clear liquor that still contains some free hydrochloric acid is sent back to the HCl absorber.

The resulting fresh acid is then ready for reuse at a fraction of the cost of replacing hydrochloric acid each time it is required. However, the sulfuric acid values contained in the by-produced copperas are definitely lost during this process and depending on the market value of the copperas produced the economics is highly dependent of the costs of sulfuric acid and geographical location.

An early version of this process appeared in 1995 in U.S. Patent 5,417,955 granted to D.W. Connolly [114] were iron was recovered at the end of the process as ferric sulfate rather than copperas, by means of air oxidation. The PHAR first appeared officially in the two patent applications to Douglas R. Olsen and Charles D. Blumenschein [115]. The U.S. Patent was finally granted in 2008 to *Olsen Environmental Management & Technologies Ltd.* (formerly Green Technology Group) [116].

Later in 2005, *Kerr-McGee Pigment GmbH* described a similar process applied to the recovery of hydrochloric acid from ferrous chloride coming from carbo-chlorination wastes [117]. In 2006, in the U.S. Patent 7,097,816 granted to Kehrmann, a variation of the PHAR was disclosed [118]. Finally, in 2008, a similar process was disclosed in the international patent application granted to the Finnish chemical company *Kemira Oy* [119].

Based on the previously mentioned active works, it exists definitely a need for recovering hydrochloric acid values from SPLs by means of a PHAR or similar process. This having bee said, there is no doubt that any process capable to recover the sulfuric acid values contained in the by-produced copperas will make the entire process even more profitable and greener.

---

[114] CONNOLLY, D.W. (1995) *Manufacture of ferric sulfate and hydrochloric acid from ferrous chloride.* U.S. Patent 5,417,955, May 23rd, 1995.

[115] OLSEN, D.R.; and BLUMENSCHEIN, C.D. (2003) *Regenerating spent pickling liquor.* U.S. Patent Applications 2002/0005210 A1, January 17th, 2002 and 2003/0026746 A1, February 6th, 2003.

[116] OLSEN, D.R.; and BLUMENSCHEIN, C.D. (2008) *System and method for converting the spent remnants of a first pickling acid solution into a usable second pickling acid solution recycling spent pickle liquor.* U.S. Patent 7,351,391, April 1st, 2008.

[117] AUER, G.; LAUBACH, B.; and VOSSING, M. (2005) *Method for treating iron chlorides and/or solutions containing iron chlorides and use of iron sulphate thus obtained.* PCT International Patent Application 2005/097682 A1, October 20th, 2005.

[118] KEHRMANN, A. (2006) *Method of producing ferrous sulfate heptahydrate.* U.S. Patent 7,097,816 B2, August 29th, 2006.

[119] MARTIKAINEN, M.; and POHJANVESI, S. (2008) *Method for recovering hydrochloric acid from iron chloride.* – PCT International Patent Application WO 2008/065258 A2 (Kemira Oy), June 5th, 2008.

---

10 Novel Sulfation Processes

10.1 Alumina from Nonbauxitic Ores

Nowadays, ninety five percent of all the metallurgical and refractory grade alumina's produced worldwide are obtained by the processing of 327 million tonnes per annum of **bauxite** [120] by the **Bayer process** that consists to perform the caustic digestion of bauxite while some alumina have been produced industrially from unconventional ores such as nepheline syenite in Russia using the lime-soda-sinter process.

Actually, the alumina content of residual bauxite ranges widely from 35 to 65 wt.% Al_2O_3 but commercial grades of bauxite require a higher and narrower alumina content of 50-55 wt.% Al_2O_3 suitable to be processed by the Bayer process. The only other sedimentary rock exhibiting a high alumina content are **aluminous clays** also called **aluminous shales** with an alumina content up to 26 wt.%.

On the other hand, the averaged alumina content of most igneous rocks both phaneritic and their microlitic equivalents (e.g., granite, rhyolite, diorite, andesite, syenite, trachyte, gabbro, basalt) is rather constant around 15.8 wt.% Al_2O_3 [121] with a peculiar exception of peraluminous igneous rocks such nepheline syenite and phonolite with up to 22 wt.% Al_2O_3 and ultramafic igneous intrusion rocks such as anorthosites exhibiting the highest alumina content from 26 wt.% to 31 wt.% Al_2O_3. Finally, the sulfation of the mineral alunite $[KAl_3(SO_4)_2(OH)_6]$ was also investigated and it will briefly described.

From the early-1970s extensive investigations and pilot testing were performed in the United States under the auspices of the former *U.S. Bureau of Mines* (USBM) and other governmental agencies and committees that were created for addressing the domestic supply of critical and strategic metals such as the *National Materials Advisory Board* (NMAB). The latter identified new extraction technologies and assessed their economic feasibility for producing metallurgical grade (M.G.) alumina or alum from nonbauxitic ores [122].

[120] LEE BRAY, E. (2018) *Bauxite and Alumina.* 2018 Minerals Yearbook. U.S. Geological Survey (USGS), Washington, DC.

[121] CLARKE, F.W. (1916) *The Data of Geochemistry, Fifth Edition.* U.S. Geological Survey, Bulletin No. 770, Washington, DC, p. 29.

[122] COLLECTIVE (1970) *Processes for Extracting Alumina from Nonbauxitic Ores.* National Materials Advisory Board (NMAB), National Academy of Sciences, National Academy of Engineering, Report NMAB-278, December 1970, Washington DC

Therefore, in the following sections, we will present the sulfation processes investigated to produce metallurgical grade alumina or aluminum sulfate directly either from anorthosites or aluminous clays.

10.1.1 Aluminum Sulfate from Anorthosites

Anorthosites are ultramafic igneous rocks made predominantly of calcic plagioclase feldspars (90%) forming solid solutions between the two end member albite ($NaAlSi_3O_8$) and anorthite ($CaAl_2Si_2O_8$) along with 10% of ferromagnesian (mafic) minerals such as pyroxene, magnetite, ilmenite and olivine.

Major occurrences of large anorthosite intrusions such as plutons and batholiths are found in all the major cratons worldwide, such as in Southwest United States (Stillwater, Montana), the Appalachian Mountains, The Grenville Province in Eastern Canada (Quebec, Ontario, and Northern Labrador), Scotland, and across Southern Scandinavia (Norway, Sweden), Russia (Kola peninsula), Eastern Europe, Northern China (Damiao Complex), and finally Korea. Several anorthosite deposits are often mined for their important metallic ore deposits: ilmenite (Lac Tio, Quebec), chromite or titano-magnetite content thus the ore beneficiation process generates huge tailings of plagioclase feldspars that are currently neither processed nor monetized but simply disposed-off.

For these reasons, anorthosites could represent abundant raw materials and thus were investigated extensively in the USA, Canada and Europe and possibly in the former USSR (now CIS) as alternative source of alumina.

The first study of anorthosites as nonconventional source of alumina was conducted by St. Clair et al. from the former the *U.S. Bureau of Mines* (USBM) [123]. Their treatment using the lime-soda sinter processing route was later revisited by Quon from *Canada Center for Mineral and Energy Technology* (CANMET) in the mid-1970s [124].

Industrially, the sulfuric acid route was patented [125] and started commercially in 1984 by *Kemira-Kemi AB* the subsidiary of Finnish company *Kemira Oy* at its plant located in Helsingborg, Sweden. The company which had a long industrial experience with the sulfation of ilmenite at its titanium sulfate plant located in Pori, Finland, for the production of titanium dioxide, decided to use a similar approach to perform the sulfation of the anorthosite concentrate. The concise flow diagram of the Kemira-Kemi process is depicted in Figure 28.

[123] St. Clair, H.W.; Elkins, D.A.; Shibler, B.K.; Mahan, W.M.; Merritt, R.C.; Howcroft, M.R.; and Hayashi, M. (1959) *Operation of Experimental plant for Producing alumina from Anorthosite*. U.S. Bureau of Mines, Bulletin No. 577, U.S. Government Printing Office, Washington DC.

[124] Quon, D.H.H. (1976) *Extraction of Alumina from Canadian and American Anorthosite by the Lime-Soda Sinter Process*. CANMET Report 76-26, Canada Center for Mineral and Energy Technology (CANMET), Energy, Mines and Resources Canada, Ottawa, ON.

[125] Jokinen, S.; and Nieminen, P. (1984) *Method for treating anorthosite and using the by-product obtained by the procedure*. Finnish Patent FI 71916 C (Kemira Oy).

The anorthosite originated from tailings from the Lapland region. After beneficiation and mineral dressing, the anorthite concentrate contained at least 30 wt.% alumina with a particle size distribution after milling, d_{50}, less than 60 μm.

The process consisted to mix the ground anorthosite concentrate with concentrated 93 wt.% H_2SO_4 inside brick-lined digesters. Afterwards water was injected into the mixture until the final concentration reached finally 50 wt.% that was accompanied by step increase of the temperature up to the boiling point of the diluted acid (130°C) due to the exothermic release of hydration enthalpy.

Actually, as explained in Section 8.7, when mixing water and 93 wt.% sulfuric acid both at ambient temperature, the specific enthalpy released by unit mass of solution is close to the maximum attainable and then the mixture reaches the highest temperature. The digestion was then conducted during two hours under atmospheric pressure and constant stirring.

Assuming for simplicity that the anorthosite is theoretically made only of the mineral anorthite when it is reacted in stoichiometric proportion with 50 wt.% sulfuric acid, the following chemical reaction occurs:

$$CaAl_2Si_2O_8 + 4H_2SO_4 + 21.8H_2O = CaSO_4.2H_2O + Al_2(SO_4)_3 + 2SiO_2 + 23.8H_2O$$

From the above chemical reaction scheme, the stoichiometric (theoretical) sulfuric acid-to-alumina mass ratio [H_2SO_4-to-Al_2O_3] is 3,840 kg (100 wt.% H_2SO_4) per tonne of alumina. Therefore, when considering the final strength during attack of 50 wt.% H_2SO_4, the sulfuric acid the practical (actual) sulfuric acid-to-solid mass ratio [A-to-S] was 2,760 kg (50 wt.% H_2SO_4) per tonne of pure anorthite.

The major reason being the rheological behavior of the suspension formed. Actually, due to the quick gelling abilities of silicates and poor filterability of the resulting jelly, it was mandatory that the liquid-to-solid mass ratio to be the highest.

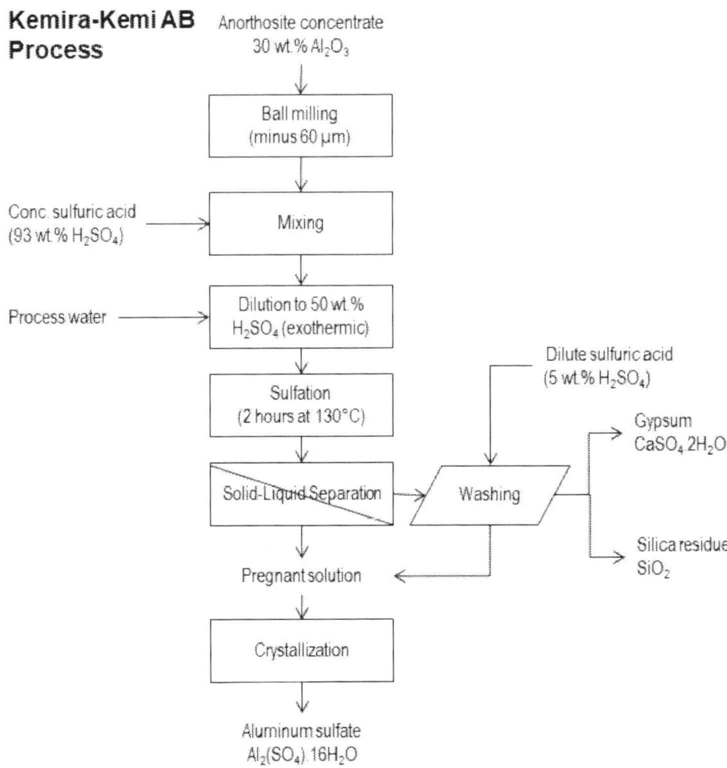

Figure 28 – Flow diagram for the Kemira-Kemi AB process

Afterwards, the suspension obtained was filtered and the gypsum ($CaSO_4.2H_2O$) along with unreacted solids and the insoluble silica residues were thoroughly washed with diluted sulfuric acid (5 wt.% H_2SO_4) to recover most of the aluminum values. Then, the filtered pregnant leach solution was used to crystallize aluminum sulfate hexadecahydrate [$Al_2(SO_4)_3.16H_2O$] that the parent company *Kemira Oy* was selling worldwide as flocculating agent along with copperas from its Pori plant for wastewater treatment. In 1984, Lunden reported in an article published in the UK trade magazine *Industrial Minerals* [126] that 40,000 tonnes of alum were produced that way and the production reached later 60,000 tonnes per annum.

[126] LUNDEN, E. (1984) Kemira-Kemi. *Industrial Minerals*. No. 7-9, pp. 540-541.

In was only in 1995, that Veldhuysen a researcher from the *Ontario Geological Survey* (OGS) [127] revisited the acidic routes using either hydrochloric acid or sulfuric acid leaching for performing the aluminum extraction process from Canadian anorthosite concentrates. Only the latter approach will be described here.

The high calcium anorthosite used in the OGS report originated from the Shawmere Archean type intrusive complex in North-Eastern Ontario. The petrographic analysis revealed the following average modal composition: 88% calcic plagioclase (75-85 An), the remaining being ferromagnesian secondary minerals such amphibole, chlorite, and garnet. The average chemical composition according to Riccio [128] is reported in the Table 58 shows rather high level of alumina suitable as an unconventional nonbauxitic ore.

The run-of-mine (R-o-M) was first beneficiated by performing the removal of ferromagnesian minerals by high intensity magnetic separation.

Table 58 – Chemical composition of the Shawmere complex anorthosite

Oxide	Mass percentage (wt.%)
SiO_2	47.70
Al_2O_3	31.10
CaO	15.00
Na_2O	2.43
Fe_2O_3	0.82
MgO	0.50

The hot sulfuric acid leaching (HAL) was performed on ground anorthite concentrate with a particle size distribution less than 100-mesh (150 μm). The process was performed batch wise using 54 wt.% H_2SO_4 at its boiling point (129°C) during 4 hours to 6 hours under atmospheric pressure.

The key figures related to the sulfation of anorthosite are summarized and reported in Table 53 hereafter.

[127] VELDHUYSEN, H. (1995) *Aluminium Extraction from an Ontario Calcic Anorthosite by Acid Processes and Resultant Products – Alumium Chemicals, Coatings, Fillers, Absorbent and Cement Additive.* Ontario Geological Survey, Open File Report 5919, Toronto, ON.

[128] RICCIO, L. (1981) *Geology of the North-eastern Portion of the Shawmere Anorthosite Complex, District of Sudbury.* Ontario Geological Survey (OGS), Open File Report 5338, Ottawa, ON, Canada.

Table 59 – Sulfation of anorthosite: key figures

ANORTHOSITE (concentrate)			
Chemical composition		Strength H_2SO_4 (initial)	**93%**
SiO_2	47.70 wt.%	Strength H_2SO_4 (attack)	54%
Al_2O_3	31.10 wt.%	$[H_2SO_4$-to-Solid](Theor.)	**1,210** kg/tonne
CaO	15.00 wt.%	$[H_2SO_4$-to-Al_2O_3](Theor.)	**3,892** kg/tonne
Na_2O	2.43 wt.%	Mass excess of acid	**0%**
MgO	0.50 wt.%	$[H_2SO_4$-to-Solid](Actual)	**1,210** kg/tonne
	96.73 wt.%	$[H_2SO_4$-to-Al_2O_3](Actual)	**3,892** kg/tonne
		[Acid-to-Solid](Actual)	**1,302** kg/tonne
		Solids pulp density	43%
		c_p (solid)	758 J/kg/K
		c_p (acid)	1467 J/kg/K
		c_p (pulp)	**2009** J/kg/K
		$\Delta h_{Sulfation}$	**-1164** kJ/kg of solid
Mass of solids (dry)	**1000** kg	$\Delta h_{Hydration}$	-185 kJ/kg of acid
Mass of sulfuric acid	1302 kg	$\Delta h_{Vaporization}$	2256 kJ/kg of steam
Mass of water (injected)	940 kg	Δh_{Total}	**284** kJ/kg of mixture
Mass of pulp	**3242** kg	Adiabatic temp rise ΔT	**-142 K ENDOTHERMIC**

Assuming for simplicity that the anorthosite is only made of the mineral anorthite when it is reacted in stoichiometric proportion with 54 wt.% sulfuric acid, the following chemical reaction occurs:

$$CaAl_2Si_2O_8 + 4H_2SO_4 + 18.55H_2O = CaSO_4.2H_2O + Al_2(SO_4)_3 + 2SiO_2 + 20.55H_2O$$

From the above chemical reaction scheme and as calculated previously, the stoichiometric (theoretical) sulfuric acid-to-alumina mass ratio $[H_2SO_4$-to-$Al_2O_3]$ is 3,840 kg (100 wt.% H_2SO_4) per tonne of alumina. However, for the Shawmere anorthosite with the composition disclosed above, the practical mass ratio of sulfuric acid-to-alumina $[H_2SO_4$-to-$Al_2O_3]$ used in the study was higher at 7,784 kg of H_2SO_4 per tonne of Al_2O_3.

Therefore, when considering the 54 wt.% sulfuric acid the practical (actual) sulfuric acid-to-solid mass ratio [A-to-S] used during the attack was very high at 2,242 kg (54 wt.% H_2SO_4) per tonne of pure anorthite concentrate.

The major reason being this choice of operating condition during sulfation was the rheological behavior of the thin suspension formed.

Actually, due to the quick gelling of silicates and poor filterability of the resulting jelly, it was mandatory that the liquid-to-solid mass ratio to be the highest with in this case a pulp density of solids of 43 mass percent.

Afterwards, the suspension obtained was filtered and the gypsum ($CaSO_4.2H_2O$) along with unreacted solids and the insoluble silica residues were thoroughly washed with diluted sulfuric acid (5 wt.% H_2SO_4) to recover most of the aluminum.

Both the clear pregnant leach solution (PLS) and the acidic wash water were combined prior performing the selective crystallization of the sodium alum [$NaAl(SO_4)_2.12H_2O$] or the directly aluminum sulfate hexadecahydrate [$Al_2(SO_4)_3.16H_2O$]. The results from the pilot scale test work confirmed a good aluminum recovery rate of nearly 85%.

Despite sulfuric acid is a cheaper chemical, exhibits a lower boiling point and lower volatility compared to hydrochloric acid, sulfuric acid is however less aggressive towards the dissolution of plagioclase feldspars. Moreover, due to the significant water of crystallization of the sodium alum or aluminum sulfate this requires significant energy to remove the water to obtain the dry alumina by calcination. Thus the study concluded that the hot acid leaching using hydrochloric acid was the most promising route for processing anorthosite.

10.1.2 Alumina from Aluminous Clays

Beside the anorthosites discussed previously, the NAMB committee also extensively investigated the possibility to process of ***aluminous clays*** for producing metallurgical grade alumina by either the alkaline route especially the lime-soda-sinter route or by the acid route.

Aluminous clays are sedimentary residual rocks that exhibit a high alumina content of 26 wt.% Al_2O_3 that were only used industrially as source of alumina in Japan during World War II.

Regarding specifically the sulfuric acid route, the comprehensive work done by Petersen and al. [129] researchers at the *U.S. Bureau of Mines* (USBM) who also investigated the processing using several other inorganic acids selected for the study (i.e., HCl, H_2SO_3, HNO_3) is remarkable. During the process, dehydrated clay was hot acid leached (HAL) counter-currently with 40 wt.% H_2SO_4 at 80°C for several hours and the pregnant leach solution (PLS) obtained was separated by gravity settling and filtration to remove the unreacted clayey material and ubiquitous colloidal silica.

---

[129] PETERS, F.A.; JOHNSON, P.W.; and KIRBY, R.C. (1963) *Methods for Producing Alumina from Clay – An Evaluation of Three Sulfuric Acid Processes*. U.S. Bureau of Mines (USBM) Report Investigation RI 6229, US Govt. Printing Office, Washington DC.

Because essentially all the iron present in the aluminous clay is often present in the ferric oxidation state, after performing the hot acid leaching, the entirety of the iron ended as iron (III) sulfate. Thus, two options were designed after the sulfation step to address the efficient removal of all the iron (III) cations: (1) The first option consisted to electrolyze the clear PLS inside an undivided horizontal electrolyzer equipped with a liquid mercury (Hg) cathode at the bottom to perform the cathode deposition of metallic iron forming an iron-mercury (Fe-Hg) amalgam a unique compound only obtained by electrolysis; and (2) the chemical removal of iron was achieved by adding manganese (II) sulfate and then oxidizing the Mn (II) to Mn (IV) sparging ozone gas that precipitated the MnO_2 entraining all the Fe (III) by co-precipitation. After concentration of the iron-depleted liquor by thermal evaporation, the selective crystallization of the aluminum sulfate occurred. The aluminum (III) sulfate obtained was thoroughly washed, oven dried and decomposed by firing to yield alpha-alumina and to regenerate the sulfuric acid.

Conversely, another alternate acidic route they investigated was to perform the above hot sulfuric acid leaching but adding potassium sulfate into the sulfuric acid solution. After leaching, the ferric sulfate was then reduced to Fe (II) by injecting sulfur dioxide gas into the pregnant leach solution and the potassium alum was crystallized leaving behind a clean ferrous sulfate liquor suitable for waste water treatment. After filtration, and thorough washing, the coarse alum crystals were dissolved into diluted potassium sulfate brine. Afterwards, the clarified alumina-bearing liquor was hydrolyzed under high pressure inside an autoclave, yielding hydrous alumina and regenerating potassium sulfate and sulfuric acid to be recycled upstream. The alumina was dried and calcined.

It is interesting to note, that the NAMB concluded in the above cited report that among the various acidic routes tested by Petersen et al., the hot acid leaching using hydrochloric acid was far more efficient than the sulfuric acid route with an operating cost close to that of the Bayer process. The major reasons being that: hydrochloric acid is more chemically reactive towards aluminum and iron oxides with faster dissolution kinetics, better decomposition of silicate minerals, and because the solubility of ferric chloride is greater than that of ferric sulfate allowing a greater removal of iron and its recovery either by pyro-hydrolysis or spray roasting is easier than calcining metal sulfates and then to convert the sulfur dioxide gas back into sulfuric acid.

The only known process implemented industrially was the so-called ***H+-Process*** invented by Maurel and Duhart and patented by *Aluminium Pechiney* in 1972 [130]. The process consisted to perform the continuous treatment of aluminous clays or schists for the recovery of alumina, iron, magnesium and alkali metals. The full technical and economic description of the entire process was

[130] MAUREL, P.; and DUHART, P. (1975) *Process for the Continuous Acid Treatment of Crude Clays and Schists.* U.S. Patent 3,862,293 (Aluminum Pechiney), January 21st, 1975.

reported in 1977 by Baudet [131] from the Department of mineralurgy of *Bureau de Recherches Géologiques et Minières* (BRGM). The flow diagram describing the process is depicted in Figure 29.

First, the aluminous clay was ground and then roasted at 700°C inside a rotary kiln in order to destroy organics and to decompose any traces of pyrite. Then the finely ground and roasted concentrate was digested with hot concentrated sulfuric acid (96 wt.%) between 125°C and 140°C during 4 hours.

After cooling, the sulfation cake obtained was then dissolved into warm water. The solids mostly silica, gypsum, and unreacted clay were removed by gravity settling and filtration and thoroughly washed with warm water to recover most of the aluminum. All the ferrous cations, if present, were oxidized into ferric cations by injecting chlorine gas (Cl_2) directly into the acidic pregnant solution according to the following chemical reaction:

$$Al_2(SO_4)_3 + 6HCl + 12H_2O = 2AlCl_3.6H_2O + 3H_2SO_4$$

Because of the high concentration of sulfuric acid remaining in solution (800-850 g/L) after the dissolution, removal of most of the impurities was achieved by precipitating the low solubility double sulfates such as: potassium and iron (III) double sulfate of chemical formula $KFe(SO_4)_2.12H_2O$.

---

[131] BAUDET, G. (1977) *A Documentary Study on Alumina Extraction Processes*. Rapport 77 SGN 061 MIN, Bureau de Recherches Géologiques et Minières (BRGM), Orleans, France.

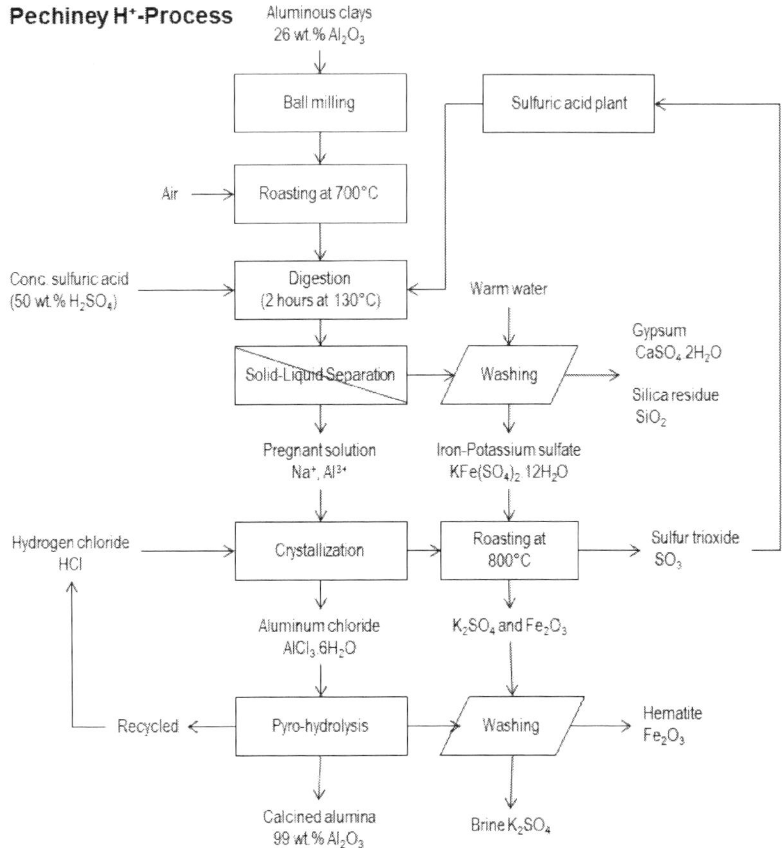

Figure 29 – Flow diagram for the Pechiney H+-Process

The aluminum values were recovered from the clear solution by precipitating hydrated aluminum chloride, $AlCl_3.6H_2O$ by sparging directly hydrogen chloride (HCl) gas. The aluminum chloride is then pyro-hydrolysed to yield pure calcined alumina and regenerates the hydrogen chloride gas recycled upstream.

The potassium and iron (III) double sulfate is roasted under controlled conditions of temperature. During roasting, the iron sulfate is selectively decomposed and yields sulfur trioxide (SO_3) gas reused upstream for the production of sulfuric acid and iron (III) oxide while the potassium sulfate remains unaffected.

After cooling, the intimate mixture of pure hematite (α-Fe_2O_3) mixed with potassium sulfate (K_2SO_4) is simply washed with warm water to produce a concentrated brine of potassium sulfate leaving behind the red iron oxide of some marketable value.

10.1.3 Alumina from Alunite

Alunite is a hydroxylated aluminium potassium sulfate mineral, with the formula $KAl_3(SO_4)_2(OH)_6$ similar to Natrojarosite.

Alunite occurs as veins and replacement masses in potassium rich volcanic rocks such as trachyte, and rhyolite. It is formed during weathering of these volcanic rocks by the action of percolating sulfuric acid formed to the oxidation of metal sulfide deposits. It is also found near volcanic fumaroles in Sicily. Historically extensive deposits were mined in Tuscany and Hungary, and New South Wales, Australia. It is currently mined at Tolfa, Italy. In the United States it is found in the San Juan district of Colorado; Goldfield, Nevada; the ghost town of Alunite, Utah; and Red Mountain, Arizona.

The sulfuric acid digestion of the mineral alunite was investigated in the late 1980s by Froisland, and al. [132] from the form *U.S. Bureau of Mines* (USBM).

10.1.4 Alum and Metallic Iron from Bauxite Residues

Red muds and bauxite residues are well-known industrial effluent slurries and solid wastes by-produced during the Bayer process consisting of the caustic digestion of bauxite with sodium hydroxide to produce metallurgical grade alumina.

Bauxite residues are mainly composed of iron (III) oxide (10-60 wt.% Fe_2O_3) imparting its characteristic reddish-brown color. The remaining major components are in decreasing order silica (5-45 wt.% SiO_2), alumina (5-25 wt.% Al_2O_3), calcia (3-15 wt.% CaO), titania (2-15 wt.% TiO_2), and residual alkalinity (10 wt.% Na_2O).

It is estimated that every tonne of metallurgical grade alumina produced, generates roughly 1.5 tonnes of red mud. Therefore, according to *International Aluminium* about 175 million tonnes of red mud were produced in 2020.

Considering that red mud landfills and dams represent a serious environmental hazard to nearby populations and the environment especially in the recent years, it is not surprising that several attempts have been made in the last decades to address this huge problem.

---

[132] FROISLAND, L.J.; WOUDEN, M.L.; and HARBUCK, D.D. (1989) *Acid Sulfation of Alunite*. U.S. Bureau of Mines (USBM) Report Investigation RI 9222, U.S. Government Printing Office, Washington DC.

Despite many approaches were made to process these wastes by either acidic or caustic approaches, the utilisation of sulfuric acid was rather limited to the conventional hot acid leaching (HAL).

For instance, recently a joint pilot scale demonstration plant was established by *Mytilineos* and *II-VI Inc.* with the aim to recover the scandium values from the pregnant solution as described in details by Balomenos et al. [133] while attempts by means of sulfation techniques *senso-stricto* involving concentrated sulfuric acid were relatively scarce.

In 2019, Anawati and Azimi from University of Toronto have experimented at the bench scale the sulfuric acid baking followed by the leaching with water of the sulfated solids with the aim to recover the scandium values [134].

The bauxite residue was mixed with concentrated sulfuric acid and baked in a furnace at 200-400°C. The mixture of sulfated solids and thermally decomposed oxides was leached with water at ambient temperature.

According to the authors, when compared with the previous hot sulfuric acid leaching processes, the devised process exhibits both reduced sulfuric acid and water consumptions, and a faster kinetics due to the high temperature involved during the sulfation.

A particular attention was made on the reaction mechanism. The operating temperature was controlling factor that dictates the final phases of the process. The findings were that performing the sulfuric acid digestion at 200°C converted most of the Fe(III) into the highly soluble phase $(H_3O)Fe(SO_4)_2$, while performing the sulfation baking at 400°C yield $Fe_2(SO_4)_3$ with a slower dissolution kinetic.

However, none of previous approaches described above address the monetization of the iron values the latter remaining mainly as sulfate of iron (III) with a low commercial value.

For that reason, since 2016, *Electrochem Technologies & Materials Inc.* conducted several prototype and pilot testing campaigns for performing the sulfuric acid digestion of red muds (RMs) and bauxite residues (BRs), originating from an alumina refinery [135].

---

[133] BALOMENOS, E.; NAZARI, G.; DAVRIS, P.; ABRENICA, G.; PILIHOU, A.; MIKELI, E.; PANIAS, D., PATKAR, S. and XU, W-Q. (2021) Scandium extraction from bauxite residue using sulfuric acid and a composite extractant-enhanced ion-exchange polymer resin. *Rare Metal Technology 2021. The Minerals, Metals & Materials Series.* Springer, Cham.

[134] ANAWATI, J., and AZIMI, G. (2019) Recovery of scandium from Canadian bauxite residue utilizing acid baking followed by water leaching. *Waste Management,* **15**(95)549-559.

[135] ONDREY, G. (2022) Recycling bauxite residues and electrowinning iron. *Chementator, Chemical Engineering magazine,* New York, NY, June 1, 2022, page 9.

The process consisted to digest the waste materials with concentrated sulfuric acid (93 wt.% H_2SO_4). From the pregnant solution obtained after dissolution of the reaction mass with warm water, gypsum was first removed by gravity settling and cross-flow filtration, thoroughly washed, and oven-dried yield a pure product.

Secondly after performing the electrochemical reduction of Fe(III) into Fe(II) followed by the crystallization, and removal of ferrous sulfate heptahydrate (copperas), the aluminum and sodium sulfates were recovered as well.

Afterwards, pure electrolytic iron flakes (99.995 wt.% Fe) were produced by electrowinning metallic iron from the copperas using the patented FerWIN™ process (see Section 11.1.6) with an OPEX of only $315 USD per tonne of electrolytic iron [136].

The electrolysis was performed inside a rectangular tank electrolyser with hanging cathodes totalizing 10 ft² (0.929 m²), anion exchange membrane compartments containing the hanging mixed metal oxides (MMO) anodes of the Ti-IrO₂ type.

During electrolysis, sulfuric acid was regenerated inside the anode compartment reaches a strength up to 30 wt.% H_2SO_4 that could be recycled upstream at the sulfation stage after conducting a thermal evaporation (see Section 12.1).

The trials confirmed the excellent faradic current efficiency (98%), throughput, and low specific energy consumption (2.9 MWh/tonne Fe) compared to the results obtained with copperas originating from other sources. The schematic of the flow diagram is depicted in Figure 30.

---

[136] CARDARELLI, F. (2009) *Electrochemical process for the recovery of metallic iron and sulfuric acid values from iron-rich sulfate wastes, mining residues, and pickling liquors.* PCT International Patent Application WO 2009/124393 (A1), October 15ᵗʰ, 2009.

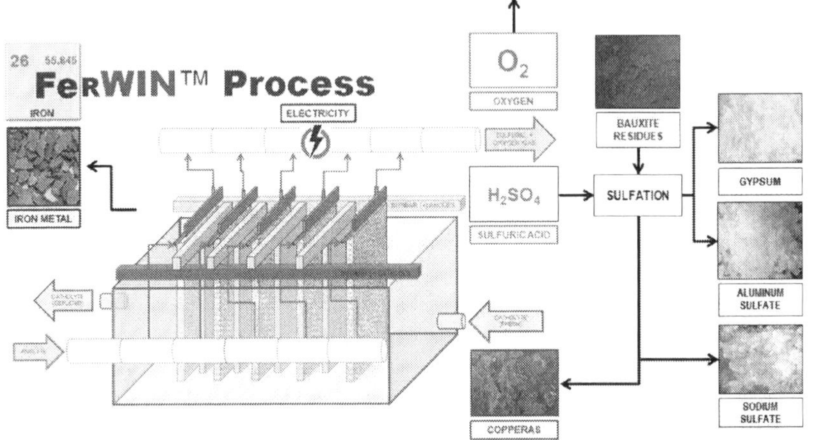

Figure 30 – Flow diagram for sulfating bauxite residues and electrowinning iron

10.2 Titanium Dioxide from Perovskite

Early 1990s, Petersen, Shirts, and Allen researchers working at the former *U.S. Bureau of Mines* (USBM), investigated a sulfation process [137] to recover pigment-grade titanium dioxide from concentrates of perovskite. **Perovskite** is a naturally occurring orthorhombic calcium titanate ($CaTiO_3$) usually disseminated as accessory mineral within basic igneous rocks such as carbonatites [138].

However a huge hard rock deposit of perovskite was discovered in Powederhorn district, Gunnison County, Colorado. The deposit named White Earth Project, was acquired by *Teck Corporation* in 1990 and consists of a carbonatite complex intruded into Precambrian granite and metamorphic rocks about 570 million years ago. The perovskite occurs in pyroxenite [139]. The resource was estimated to contain 5.9 million tonnes of TiO_2 equivalent and the run-of-mine produced contained 13.2 wt.% TiO_2 together with rare earth oxides (2 wt.% REOs) and niobium pentoxide (0.3 wt.% Nb_2O_5) with minor amount of thorium dioxide (ThO_2).

[137] PETERSEN, A.E.; SHIRTS, M.B.; and ALLEN, J.P. (1992) *Production of Titanium Dioxide Pigment from Perovskite Concentrates, Acid Sulfation Method.* Report of Investigation RI-9397 with Appendix on Economic and Technical Evaluation by J.H. Schwier, U.S. Bureau of Mines (USBM), U.S. Dept. of Interior, Washington, DC.

[138] GAINES, R.; SKINNER, H.C.W; FOORD, E.E.; MASON, B.; and ROSENWEIG, A. (1997) *Dana's New Mineralogy, Eighth Edition.* John Wiley & Sons, Inc., New York, NY, pp. 223.

[139] SHAVER, K.C.; and LUCENFORD, R.A. (1998) White Earth Project, Colorado. The largest titanium resource in the United States. CIM Bulletin **91**(1022).

At that time, the discovery of this peculiar domestic perovskite deposits was considered by the *U.S. Bureau of Mines* a significant, but still untapped, titanium resource that could potentially reduce the dependence of the United States on imports of titanium feedstocks while benefiting from additional rare earth and niobium co-products. The former if extracted was intended to be shipped to the Mountain Pass facilities, California for further processing and refining.

The sulfuric acid digestion process consisted first to perform continuously the sulfuric acid digestion of the ground perovskite concentrate (-100 mesh) at 200°C for two hours under intense mixing (1000 rev/min) to keep the solids suspended. The process was developed and tested both at the laboratory and prototype scales. The stoichiometric (theoretical) sulfuric acid-to-solid mass ratio [H_2SO_4-to-S] is 1311 kg per tonne of perovskite. But in practice, an excess of 1130% concentrated sulfuric acid (80 wt.% H_2SO_4) was used. The latter corresponds to a practical (actual) sulfuric acid-to-solid mass ratio [A-to-S] of 20,158 kg (80 wt.% H_2SO_4) per tonne of concentrate.

The technical reason behind the utilization of such large mass excess of sulfuric acid with an optimum acid-to-solid ratio of 20:1 (i.e., pulp density of 4.8 wt.% solids) was imposed by the strong chemical reactivity of the inherent perovskite concentrate due to its elevate calcium oxide content (CaO).

Thus the large amount of heat released by the exothermic sulfation reaction was entirely absorbed by the sensible heat of the sulfuric raising the temperature of the mixture near its boiling point and thus ensuring a safe autogenous mode of operation and avoiding a thermal runaway to occur.

The key figures related to the sulfation of perovskite are summarized and reported in Table 60 hereafter.

Table 60 – Sulfation of perovskite: key figures

PEROVSKITE (concentrate)			
Chemical composition		Strength H_2SO_4 (initial)	**80%**
TiO_2	50.00 wt.%	Strength H_2SO_4 (attack)	80%
CaO	33.00 wt.%	[H_2SO_4-to-Solid](Theor.)	1,311 kg/tonne
$SiO_2{}^r$	3.70 wt.%	Mass excess of acid	**1130%**
FeO	2.70 wt.%	[H_2SO_4-to-Solid](Actual)	**16,127** kg/tonne
MgO	2.30 wt.%	[H_2SO_4-to-TiO_2](Actual)	**32,254** kg/tonne
Ln_2O_3	3.00 wt.%	[Acid-to-Solid](Actual)	**20,158** kg/tonne
	94.70 wt.%	Solids pulp density	4.7%
		c_p (solid)	669 J/kg/K
		c_p (acid)	1614 J/kg/K
		c_p (pulp)	**1364** J/kg/K
		$\Delta h\ _{Sulfation}$	**-2875** kJ/kg of solid
Mass of solids (dry)	**1000** kg	$\Delta h\ _{Hydration}$	0 kJ/kg of acid
Mass of sulfuric acid	20158 kg	$\Delta h\ _{Vaporization}$	2256 kJ/kg of steam
Mass of water (injected)	0 kg	$\Delta h\ _{Total}$	**294** kJ/kg of mixture
Mass of pulp	**21158** kg	Adiabatic temp rise ΔT	**-215 K ENDOTHERMIC**

The schematic flow-diagram for recovering titanium as pigment-grade titanium dioxide from perovskite is depicted in Figure 31.

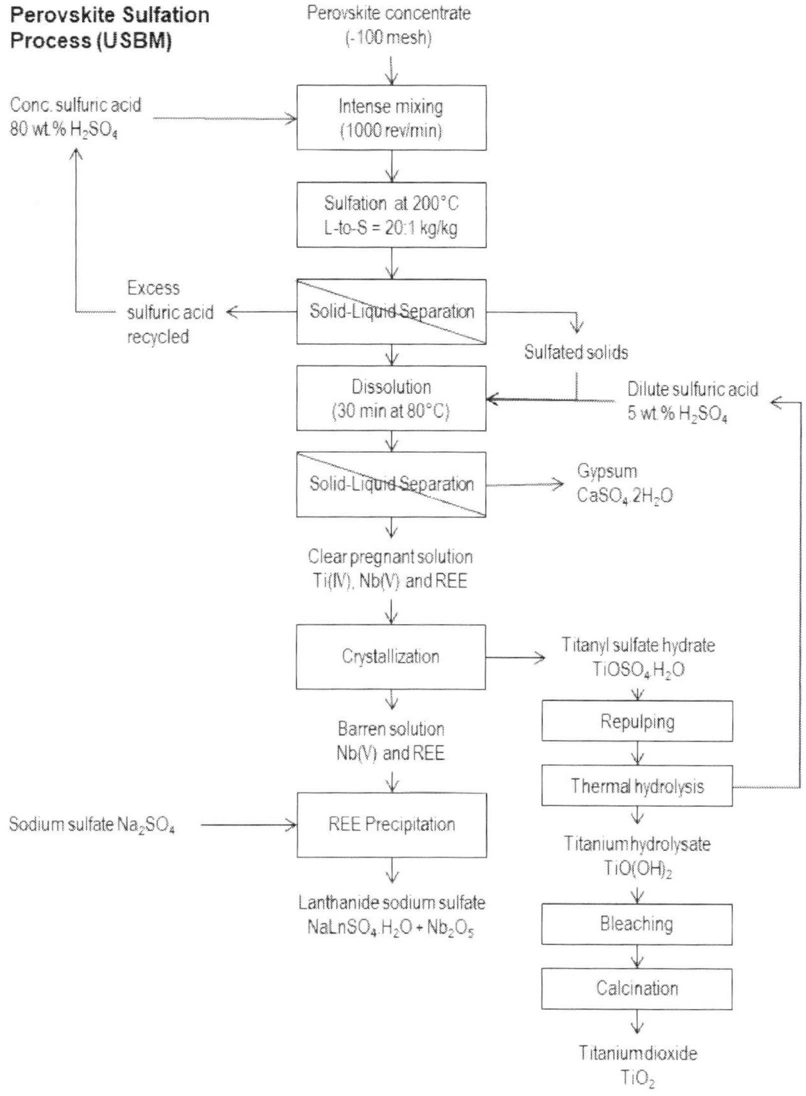

Figure 31 – Flow diagram for the USBM process

The near boiling point conditions were essential to drive-off moisture and by-produced water as vapor thus preventing further dilution with loss in titanium recovery yield.

Upon completion of the digestion and subsequent cooling, the excess of concentrated sulfuric acid was recovered by cross-flow filtration. The remaining insoluble solid residue with around 10% titanium and 0.5% lanthanides, containing coarser unreacted perovskite, and metal sulfates that are insoluble in strong sulfuric acid, was thoroughly washed with hot recycled dilute sulfuric acid solution at 80°C during 30 minutes to dissolve the titanium, niobium, and rare earth values. The gypsum along with solid residues is separated by gravity settling and filtration.

The clear pregnant solution was evaporated to crystallize titanyl sulfate monohydrate ($TiOSO_4.H_2O$) recovered by cross flow filtration while the barren solution containing sulfuric acid leftover and the lanthanide and niobium species was further regenerated to recover the free sulfuric acid together with rare earth sulfates by adding sodium sulfate, and precipitating the double salts and about fifty percent of the niobium pentoxide. The dissolution of the fresh $TiOSO_4.H_2O$ in acidic water was followed by the bleaching of the pregnant solution by reducing all the ferric iron into ferrous iron using metallic iron powder. Afterwards, the precipitation of TiO_2 by thermal hydrolysis was performed and the white precipitate washed and calcined to yield pure TiO_2. Because of the elevate calcium content, the processing of perovskite by sulfation also yields large amounts of either anhydrous calcium sulfate ($CaSO_4$) or gypsum $CaSO_4.2H_2O$ as by-products that must be pure enough to be commercialized in order to sustain the economic viability of the process.

The first three unit operations have been demonstrated using laboratory-scale continuous methods of operation: (1) sulfuric acid sulfation of perovskite concentrates, (2) dissolution of sulfation solid residues, and (3) crystallization of titanyl sulfate. Afterwards, the batch testing of other unit operations was performed: (4) re-dissolution of $TiOSO_4.H_2O$ followed by (5) bleaching of the pregnant solution using iron reduction, (6) precipitation of TiO_2 by hydrolysis, and finally (7) regenerating the spent sulfuric acid with rare-earth by-product recovery

At the end of the campaign 97 percent of both titanium and niobium values were recovered while 70 to 90 percent of the rare-earth by-products were extracted.

10.3 Titanium Dioxide from Low Grade Titanium Slags

The "Highveld Process" is a well-known steelmaking process that consists in the smelting/slagging of the vanadiferous titano-magnetite (VTM) followed by soda ash roasting to recover vanadium pentoxide. More precisely, the VTM concentrate is pre-reduced using a rotary kiln fed with pulverized coal.

The pre-reduced material is then smelted inside a submerged AC arc furnace ("slagging process") with carbon electrodes and anthracite coal to produce pig iron with 3.5 wt. % C and about 1.2 wt. % V.

The tapped primary titanium slag contains only 32 wt. % TiO_2, 15 wt.% CaO, 10 wt.%MgO, and 0.75 wt. % V_2O_5. Because it exhibits a much lower grade than commercial titanium slags such as those produced by Rio Tinto in Sorel-Tracy (Sulfate slag with 80 wt.% TiO_2), Canada, and Richards Bay (Chloride slag with 91 wt.% TiO_2), South Africa, they are not used commercially and simply landfilled.

The smelter gas (i.e., a mixture of CO and H_2) is typically reused elsewhere in the plant, but in fine the combustible gas ends up burned releasing CO_2 into the atmosphere.

The hot metal is subsequently transferred molten inside shaking ladles to a steel making plant and converted into steel inside a basic oxygen furnace (BOF) where oxygen gas is injected together with suitable fluxes to decarburize and remove the vanadium from the molten metal.

A low carbon steel of excellent commercial value and a vanadium-rich slag containing up to 20-25 wt.% V_2O_5 are thus obtained. However, the steel making process is rather energy intensive as the specific energy consumption for producing a slab of steel including smelting, refining and hot rolling is on average 6.38 MWh per tonne of steel slab. Moreover, the associated greenhouse gas emissions (GHGs) are on average 1.8 tonnes of carbon dioxide [140].

The vanadium rich slag, eventually diluted with some smelter slag, is subsequently subjected to soda ash and/or salt roasting inside a brick-lined rotary kiln with direct heating using pulverized coal.

The hot clinker is discharged from the kiln and quenched with water to yield a pregnant leach solution (PLS). The remaining steps resemble the "Vametco process" in that a desilication using magnesium and aluminum sulfates is performed, followed by the precipitation of vanadium as ammonium metavanadate (NH_4VO_3) by adding ammonium sulfate, and calcination to yield vanadium pentoxide.

To address the recycling of the primary titanium slag with the monetization of both titanium and vanadium values, a sulfation process invented by Becker and Dutton has been patented in the early 2000 by *Evraz Highveld Steel and Vanadium Ltd* in South Africa [141] specifically to address the recovery of titanium dioxide from the primary and low grade titanium slag.

The schematic flow diagram of the Highveld process as described in the above patent is depicted in Figure 32. And a brief description of the proposed process is provided hereafter.

---

[140] FRUEHAN, R.J.; FORTONI, O.; PAXTON, H. W.; and BRINDLE, R. (2000) *Theoretical Minimum Energies to Produce Steel for Selected Conditions.* Carnegie Mellon University, Pittsburgh, Pa.

[141] BECKER, J.H.; and DUTTON, D.F. (2008) Recovery of titanium dioxide from titanium oxide bearing materials like steelmaking slags. U.S. Patent 7,462,337 B2 (Evraz), December 9th, 2008.

The oven-dried primary low grade titanium-bearing slag produced during the Highveld Steelmaking process containing about 22 to 32 wt.% of titanium dioxide with a composition reported in Table 61 is ground inside a ball mill with an appropriate particle size until at least 80% passing through a 175 micrometers mesh and preferentially below 45 micrometers.

The key figures related to the sulfation of the primary titanium slag are summarized and reported in Table 61 hereafter.

Table 61 – Sulfation of Highveld titanium slag: key figures

HIGHVELD TITANIUM SLAG			
Chemical composition		Strength H_2SO_4 (initial)	96%
TiO_2	24.50 wt.%	Strength H_2SO_4 (attack)	96%
SiO_2	25.00 wt.%	[H_2SO_4-to-Solid](Theor.)	1,391 kg/tonne
CaO	15.00 wt.%	Mass excess of acid	5%
MgO	12.00 wt.%	[H_2SO_4-to-Solid](Actual)	1,461 kg/tonne
Al_2O_3	13.00 wt.%	[Acid-to-Solid](Actual)	1,522 kg/tonne
Fe_2O_3	5.00 wt.%	Solids pulp density	40%
FeO	4.00 wt.%	c_p (solid)	757 J/kg/K
V_2O_5	0.75 wt.%	c_p (acid)	1437 J/kg/K
	99.25 wt.%	c_p (pulp)	1144 J/kg/K
		Δh Sulfation	-1924 kJ/kg of solid
Mass of solids (dry)	1000 kg	Δh Hydration	0 kJ/kg of acid
Mass of sulfuric acid	1522 kg	Δh Vaporization	2256 kJ/kg of steam
Mass of water (injected)	0 kg	Δh Total	-709 kJ/kg of mixture
Mass of pulp	2522 kg	Adiabatic temp rise ΔT	619 K EXOTHERMIC

In the description of the invention, the ground titanium slag is then reacted with cold concentrated sulfuric acid (93-98 wt.% H_2SO_4) inside a continuous or batch digester.

In order to carry out the sulfation reaction, the stoichiometric (theoretical) sulfuric acid-to-solid mass ratio [H_2SO_4-to-S] is only 1,391 kg (100 wt.% H_2SO_4) per tonne of slag and it is reasonable that in practice at least a 5 mass percent excess of sulfuric acid is used.

Subsequently, still, preheated air at about 400°C is then introduced through the bottom of the reactor and allowed to rise through the reaction mix in order to mix to the point where the sulfation reaction commences at around 100°C until it reaches a plateau of 180°C.

The heated air velocity is increased and allowed to pass through the cake material in order to produce a porous cake material with final baking temperature ranging from 190°C and 250°C for about two to three hours.

However, based on the reported sulfuric acid-to-slag mass ratio, and based on the above predicted sulfation key figures, it seems difficult not to say unattainable to perform such sulfation reaction maintaining such low temperature, for the reasons explained hereafter.

Actually, due to the relatively high calcium and magnesium oxides content of the slag in some instance totalizing 27 wt.% (MgO + CaO), the sulfation reaction of such finely ground slag is extremely exothermic (-1924 kJ/kg) as confirmed by the huge adiabatic temperature rise (619 K). It will be accompanied by a violent evolution of copious white fumes of SO_2/SO_3 coming off the reaction mixture in a matter of seconds.

Based on the author own opinion, and his practical experience having performed several trials at the prototype scale with other types of high calcium steelmaking slags, this scenario can be only be achieved safely without leading to a catastrophic thermal runaway by utilizing a much larger sulfuric acid-to-solid mass ratio (e.g., 200% excess sulfuric acid) than the mass ratios claimed in the patent as exemplified from the intentionally modified key figures reported in Table 62.

Table 62 – Sulfation of Highveld titanium slag (Modified): key figures

HIGHVELD TITANIUM SLAG (MODIFIED)			
Chemical composition		Strength H_2SO_4 (initial)	**96%**
TiO_2	24.50 wt.%	Strength H_2SO_4 (attack)	96%
SiO_2	25.00 wt.%	[H_2SO_4-to-Solid](Theor.)	**1,391** kg/tonne
CaO	15.00 wt.%	Mass excess of acid	**200%**
MgO	12.00 wt.%	[H_2SO_4-to-Solid](Actual)	**4,173** kg/tonne
Al_2O_3	13.00 wt.%	[Acid-to-Solid](Actual)	**4,347** kg/tonne
Fe_2O_3	5.00 wt.%	Solids pulp density	19%
FeO	4.00 wt.%	c_p (solid)	757 J/kg/K
V_2O_5	0.75 wt.%	c_p (acid)	1437 J/kg/K
	99.25 wt.%	c_p (pulp)	**1279** J/kg/K
		Δh Sulfation	**-1924** kJ/kg of solid
Mass of solids (dry)	**1000** kg	Δh Hydration	0 kJ/kg of acid
Mass of sulfuric acid	4347 kg	Δh Vaporization	2256 kJ/kg of steam
Mass of water (injected)	0 kg	Δh Total	**-287** kJ/kg of mixture
Mass of pulp	**5347** kg	Adiabatic temp rise ΔT	**224 K EXOTHERMIC**

Moreover, the finely ground material must be feed incrementally to the large excess of sulfuric acid under intense mixing. This in order to absorb the heat released progressively taking advantage of the sensible heat of concentrated sulfuric acid.

This latter approach was utilized initially by the former *U.S. Bureau of Mines* for sulfating perovskite (Section 10.2) and used recently for sulfating BOF-slags by the New Zealand Company *Avertana* and this novel technology will be described in details in Section 10.4.

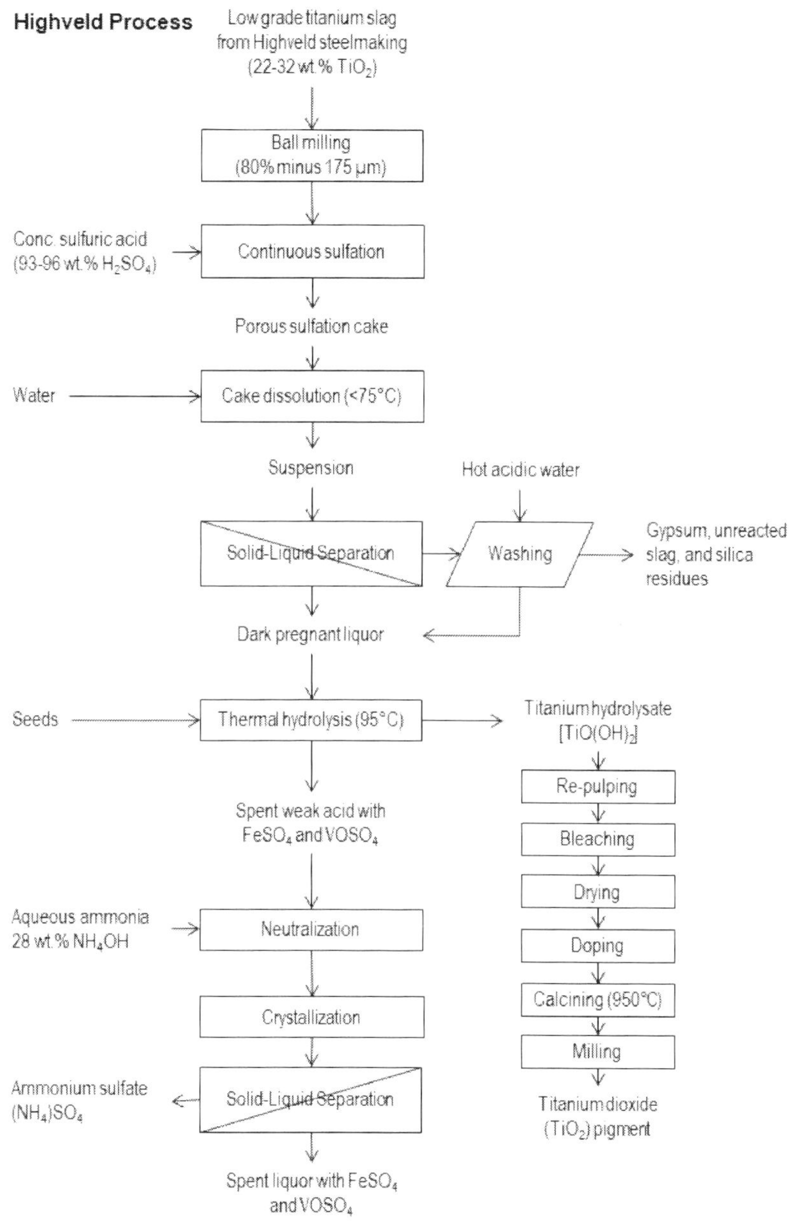

Figure 32 – Flow diagram for the Highveld process

After the sulfation reaction has proceeded substantially to completion, a dense cake containing titanyl sulfate, gypsum, and aluminum sulfate is formed.

Then, water is introduced from the bottom of the digester and allowed to flow through the cake to begin digestion with a water-to-sulfation cake mass ratio [H_2O-to-C] of 1,300 kg per tonne of cake.

During this water dissolution step, cool air is introduce into the reactor to keep the temperature below about 85°C, preferably below 75°C, depending on the feed stock (e.g., high chromium content) to avoid premature precipitation of titanium oxyhydrate. The dissolution process during which air is allowed to pass through the suspension and act as a mixing agent is continued until the cake material has been completely digested. Typically, this stage of the process takes about 4 hours until completion.

Once the cake material has been digested, the reactor is drained and after the metered addition of flocculating agents, a solid-liquid separation step is performed in order to remove gypsum, silica, and other unreacted slag residue, that are washed and then simply disposed as tailings.

The dense pregnant solution containing high concentrations of $TiOSO_4$, Al_2SO_4, $Fe_2(SO_4)_3$, $FeSO_4$, and $VOSO_4$ undergoes further processing. The mother liquor undergoes the removal of aluminum by simply adding ammonium sulfate a by-product from the downstream roasting plant that crystallizes upon cooling as bulky ammonium alum crystals [$NH_4Al(SO_4)_2.12H_2O$].

Afterwards, traces of Ti(III) in the dense solution are oxidized with hydrogen peroxide to convert all the titanium to Ti(IV). Then a dilute suspension of rutile nuclei in water heated to 60°C is added quickly to the hot liquor at a temperature near the boiling point (95°C). That triggers the thermal hydrolysis of titanyl cations that precipitate as white insoluble $TiO(OH)_2$.

The titanium hydrolysate can be then bleached, washed, filtered, dried, etc. according to protocols well-known in the titanium dioxide pigment industry and already described in Section 9.1. The remaining spent acidic liquor is pale blue in color due to the presences of remaining Fe(II) and V(IV) cations which can be recovered eventually from the liquor if necessary.

10.4 Magnesium and Titanium from BOF-Slags

Recently in 2016, the New Zealand Company *Avertana Limited* patented a novel sulfation technology aimed to process and titanium and magnesium-rich metallurgical slags originating from steelmaking [142, 143].

[142] HASSELL, D., OBERN, J.K, MOLLOY, S.D.J., IBRAHIM, S.O.Z.E.M.; AND ALI, M.S. (2016) *Extraction of products from titanium-bearing minerals.* PCT International Patent Application WO 2016/007020 A1, January 14th, 2016.

[143] HASSELL, D., OBERN, J.K, MOLLOY, S.D.J., IBRAHIM, S.O.Z.E.M.; AND ALI, M.S. (2019) *Extraction of products from titanium-bearing minerals.* U.S. Patent 10,294,117 B2 (Avertana), May 21st, 2019.

Because of the high alkalinity imparted by the presence of calcia and magnesia, as discussed in the previous Section 10.3, the sulfation approach needs to be similar to that of the *U.S. Bureau of Mines* for sulfating perovskite and it consists to add the slag incrementally to a very large volume of concentrated sulfuric acid in order to absorb the large amount of heat released while raising the temperature of the mixture.

The schematic flow diagram of the Avertana process is depicted in Figure 33. The process consists first to ground the titanium-bearing steelmaking slag until reaching an average particle size of less than 180 μm. Then the ground material is added incrementally to a large volume of concentrated sulfuric acid (80-90 wt.% H_2SO_4) with a stoichiometric (theoretical) sulfuric acid-to-solid mass ratio [H_2SO_4-to-S] of 1, 529 kg (100 wt.% H_2SO_4) per tonne.

The sulfation reaction is conducted continuously inside a reactor, such method is mandatory to absorb the large amount of heat due to the high calcia (CaO) and magnesia (MgO) content of the slag and thus avoiding a catastrophic thermal runaway due to the strong exothermic reaction and violent evolution of noxious gases. The baking is conducted during two hours at a temperature ranging from 180°C to 210°C.

The key figures related to the sulfation of BOF steelmaking slags are summarized and reported in Table 63 hereafter.

Table 63 – Sulfation of BOF slag: key figures

BOF-SLAG			
Chemical composition		Strength H_2SO_4 (initial)	**93%**
CaO	45.00 wt.%	Strength H_2SO_4 (attack)	93%
Fe_2O_3	12.00 wt.%	[H_2SO_4-to-Solid](Theor.)	**1,529** kg/tonne
SiO_2	15.00 wt.%	Mass excess of acid	**200%**
FeO	6.00 wt.%	[H_2SO_4-to-Solid](Actual)	**4,588** kg/tonne
MnO	10.00 wt.%	[Acid-to-Solid](Actual)	**4,934** kg/tonne
MgO	10.00 wt.%	Solids pulp density	17%
Al_2O_3	2.00 wt.%	c_p (solid)	741 J/kg/K
		c_p (acid)	1467 J/kg/K
	100.00 wt.%	c_p (pulp)	**1288** J/kg/K
		Δh Sulfation	**-3050** kJ/kg of solid
Mass of solids (dry)	**1000** kg	Δh Hydration	0 kJ/kg of acid
Mass of sulfuric acid	4934 kg	Δh Vaporization	2256 kJ/kg of steam
Mass of water (injected)	0 kg	Δh Total	**-383** kJ/kg of mixture
Mass of pulp	**5934** kg	Adiabatic temp rise ΔT	**297 K EXOTHERMIC**

After sulfation, the resulting sulfation cake is dissolved into dilute sulfuric acid to form a sulfated suspension. The unreacted solids, gypsum, and silica residues are removed by cross flow filtration.

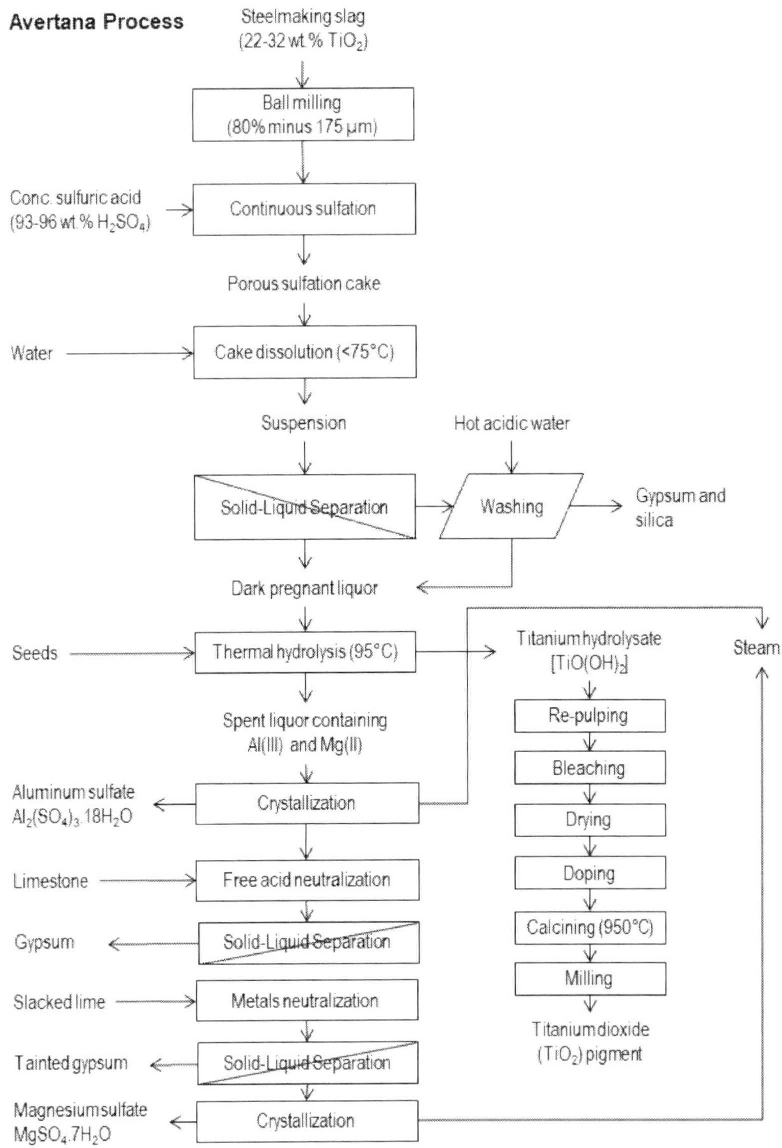

Figure 33 – Flow diagram for the Avertana process

The first clear pregnant solution containing the titanyl sulfate (TiOSO$_4$), the aluminum and magnesium sulfates undergoes first a thermal hydrolysis to produce a titanium hydrolysate [TiO(OH)$_2$] which is recovered by filtration and sent to a finishing section.

Afterwards, the second clear solution containing the aluminum and magnesium sulfates is concentrated in order to crystallize aluminium sulfate dodecahydrate [$Al_2(SO_4)_3.18H_2O$].

After removal of the crystals of aluminum sulfate by decantation the third liquor still containing free sulfuric acid is neutralized using ground limestone (i.e., calcium carbonate). The precipitated gypsum ($CaSO_4.2H_2O$) is simply removed by gravity settling and filtration.

Then slacked lime, $Ca(OH)_2$, is added to the fourth filtrate to increase the pH above 12 for removing traces of deleterious metals (e.g., Fe, Cr, Mn, V, etc.) as insoluble metal hydroxides that are co-precipitated along with gypsum yielding a brown sludge. After gravity settling the tainted gypsum solids are disposed-off and landfilled or used as cementitious product in the cement manufacture [144].

The neutral fourth liquor is then concentrated in order to crystallize pure Epsom's salt or magnesium sulfate heptahydrate ($MgSO_4.7H_2O$).

10.5 Potassium Sulfate from Greensands

Potassium (potash) together with phosphorus already discussed in Section 9.2 plays a vital role in supplying key nutrients to the top soil to support intensive crop production. Presently, K-fertilizers are essentially produced from sedimentary evaporitic deposits containing the following halides minerals: ***sylvinite, syngenite, carnalite, kainite, polyhalite***, and surface and sub-surface brines. The total potassium content in potash fertilizer is expressed as mass percent of potassium oxide (K_2O).

In 2021, the major potash producing countries are in order of importance [1 Mt = 10^6 tonnes (E)]: Canada (14 Mt), Russia (9 Mt), Belarus (8 Mt), China (6 Mt), Germany (3 Mt), Israel (2.3 Mt), Jordan (1.6 Mt), Chile (0.9 Mt), United States (0.48 Mt) and finally Spain (0.40 Mt).

Since the 1900s, greensands that contain the clay mineral called ***glauconite*** which is a natural occurring iron-rich, heterogeneous, phyllosilicate mineral contains around 4-8 wt.% of potash (K_2O) have been considered as an alternative in locations where evaporitic deposits were lacking from a geological standpoint. Despite natural greensands have been used directly as fertilizer, several attempts have been made to extract the potash.

Commercial production of potash from greensands using the ***Shreve Process*** started in the 19th century in the USA by the *Eastern Potash Corp.* and consisted to perform the roasting of beneficiated green sands with quicklime.

---

[144] OBREN, J. (2021) *Cement Additive*. PCT International Patent Application WO 2021/125979A1, June 24th, 2021.

Table 64 – Average chemical composition of greensand

Oxide	Mass percentage (wt.%)
SiO_2	50.5
Fe_2O_3	17.5
Al_2O_3	5.8
K_2O	7.1
MgO	3.8
FeO	3.1
CaO	0.9
Na_2O	0.4
LOI	9.0

Later, experimental attempts were made at the laboratory scale to solubilise the potassium values as potassium sulfate by sulfating directly the greensand as devised by Turrentine et al. in 1921 [145] a similar process was revisited recently by Shekhar et al. [146].

The only sulfation process that was commercialized was called the **Moxham process** named after A.J. Moxham president of the *Electro Company*.

The Moxham process consisted to perform the sulfation of greensands with the chemical composition close to that reported in Table 64 with hot concentrated sulfuric acid (60 wt.% H_2SO_4) produced from the Chamber process.

The heat necessary to reach the temperature of 90°C necessary to start the reaction was simply provided by diluting the sulfuric acid with water. Based on the above composition, the stoichiometric (theoretical) sulfuric acid-to-solid ratio [H_2SO_4-to-S] is 721 kg (100 wt.% H_2SO_4) per tonne of greensands but the practical sulfuric acid-to-solid ratio [A-to-S] used in the process was close to 1,201 kg (60 wt.% H_2SO_4) per tonne. After 5 to 6 hours of digestion the sulfation was complete.

The key figures related to the sulfation of greensands are summarized and reported in Table 65 hereafter.

[145] TURRENTINE, J.W.; WHITTAKER, C.W.; and FOX, E.J. (1925) Potash from greensand (Glauconite). *Industrial and Engineering Chemistry*, **17**(11)1177-1181.

[146] SHEKHAR, S.; MISHRA, D.; AGRAWAL, A.; AND SAHUB, K.K. (2017) Physical and chemical characterization and recovery of potash fertilizer from glauconitic clay for agricultural application. *Applied Clay Science*, **143**, 50-56.

Table 65 – Sulfation of greensands: key figures

GREENSANDS (concentrate)			
Chemical composition		Strength H_2SO_4 (initial)	60%
SiO_2	50.50 wt.%	Strength H_2SO_4 (attack)	60%
Fe_2O_3	17.50 wt.%	[H_2SO_4-to-Solid](Theor.)	721 kg/tonne
K_2O	7.10 wt.%	Mass excess of acid	0%
Al_2O_3	5.80 wt.%	[H_2SO_4-to-Solid](Actual)	721 kg/tonne
MgO	3.80 wt.%	[H_2SO_4-to-K_2O](Actual)	10,150 kg/tonne
FeO	3.10 wt.%	[Acid-to-Solid](Actual)	1,201 kg/tonne
CaO	0.90 wt.%	Solids pulp density	45%
Na_2O	0.40 wt.%	c_p (solid)	649 J/kg/K
H_2O	9.00 wt.%	c_p (acid)	1907 J/kg/K
	98.10 wt.%	c_p (pulp)	1058 J/kg/K
		Δh $_{Sulfation}$	-787 kJ/kg of solid
Mass of solids (dry)	1000 kg	Δh $_{Hydration}$	0 kJ/kg of acid
Mass of sulfuric acid	1201 kg	Δh $_{Vaporization}$	2256 kJ/kg of steam
Mass of water (injected)	0 kg	Δh $_{Total}$	135 kJ/kg of mixture
Mass of pulp	2201 kg	Adiabatic temp rise ΔT	-127 K ENDOTHERMIC

After cooling, the siliceous solid residue was separated by filtration from the liquor and then leached (bleaching) with hydrochloric acid to remove the traces of iron (III). After washing with hot water, the whitish silica residue was dried to yield a suitable absorbent of commercial value.

The clear pregnant leach solution that contained all the sulfates of aluminum, potassium, and iron was processed in one-step. Actually, sulfuric acid was added to the solution to crystallize the metal sulfates all at once. After separating the crystalized mass by filtration, the spent acidic solution containing some sulfuric acid was recycled.

The granular mixture of crystals of potassium alum, copperas, and ferric sulfate was dried and then roasted at 500°C inside a rotary kiln. At that temperature, all the iron sulfates decompose yielding sulfur trioxide that was reacted with the spent acidic solution to regenerate at least a portion of the sulfuric acid.

After dissolving the roasted mass with hot water, a high quality iron (III) oxide pigment (α-Fe_2O_3) was produced while the solution contained all the potassium and aluminum that crystallized readily to **K-alum** of chemical formula $KAl(SO_4)_2.12H_2O$.

Then the K-alum was then roasted at 900°C to yield pure calcined alumina (Al_2O_3), and potassium sulfate (K_2SO_4) and evolving sulfur trioxide. After dissolution with hot water, the alumina was recovered, and the brine containing potassium sulfate was concentrated by evaporation.

10.6 Niobium and Rare Earths from Pyrochlore

Since the early 2010, *Eramet SA* developed a new hydrometallurgical process to recover niobium, rare earth elements, and uranium from the Mabounié deposit in Gabon. Located near Lambarené in central Gabon about 200 km from Libreville, the world-class Mabounié polymetallic ore deposit discovered in 1986 by the *Bureau de Recherches Géologiques et Minières* (BRGM), has, since 2005, been developed by Maboumine, a company owned partly by *Compagnie Minière de l'Ogooué* (Comilog) (76%), the Gabonese state (15%) and other minority shareholders (9%), under the leadership of *Eramet SA*.

The weathered carbonatite of the Mabounié alkaline complex is hosting the pyrochlore but exhibits a rather complicated chemistry and mineralogy [147] compared to other world class niobium deposits (e.g., Araxa, Lueshe, and Saint-Honoré) as the niobium is contained mainly in the pyrochlore and ferrocolumbite mineral phases, while the rare earths are hosted in both pyrochlore, apatite, and the mineral ***crandallite*** $[CaAl_3(PO_4)(PO_3OH)(OH)_6]$.

The hydrometallurgical process to address the treatment of the above pyrochlore was invented by Agin et al. and patented by *Eramet* in 2012 [148]. A schematic of the flow diagram is depicted in Figure 34.

---

[147] LAVAL, M.; JOHAN, V.; and TOURLIERE, B. (1988) The Mabounie carbonatite: An example of the formation of a residual deposit with pyrochlore. *Chronique de la Recherche Minière* **56**, 125-136

[148] AGIN, J.; DURUPT, N.; GRECO, A.; HAMMY, F.; LAROCHE, G.; and THIRY, J. (2012) *Dissolution and recovery of at least one element nb or ta and of at least one other element u or rare earth elements from ores and concentrates.* PCT International patent Application, WO 2012/093170A1 (Eramet), July 12th, 2012.

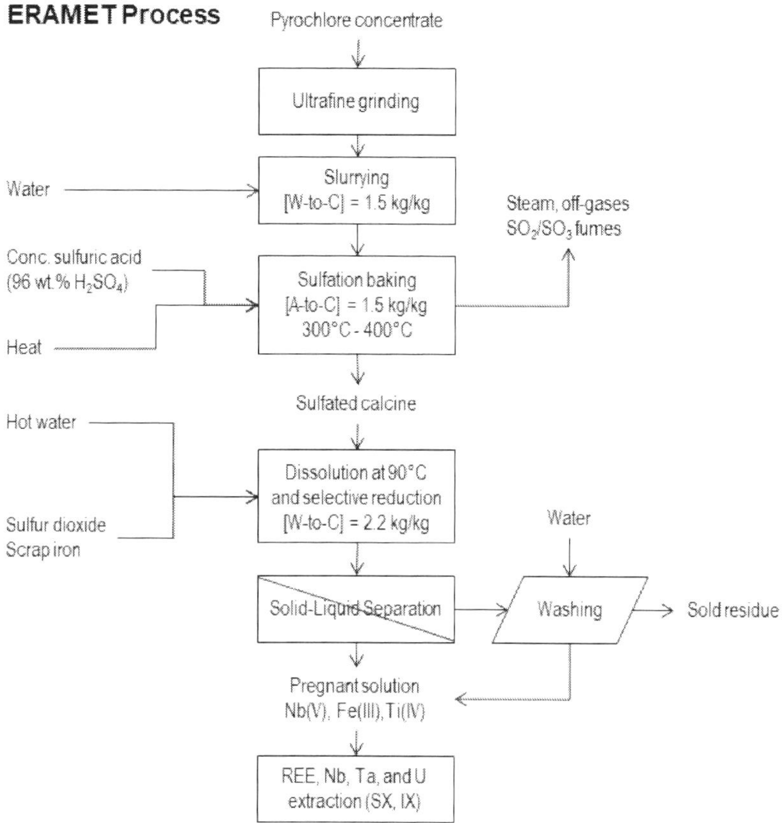

Figure 34 – Flow diagram for the Eramet process

Several flow sheets are described in the patent but on average the process consists to perform first the ultrafine grinding of the pyrochlore with a particle size distribution (80%) ranging from 4 to 30 micrometers to enhance the chemical reactivity during sulfation.

The milled concentrate is then slurried with water to obtain a suspension with a pulp density ranging from 35 wt.% to 45 wt.%. Then concentrated sulfuric acid (96 wt.% H_2SO_4) is added and vigorously mixed to ensure an homogeneous paste. Alternatively, it is claimed that sulfur trioxide can be injected into the slurry to produce *in-situ* a fraction or all the sulfuric acid required. The average sulfuric acid-to-concentrate mass ratio [A-to-S] is roughly 1,500 kg per tonne.

The mixture is then roasted inside a kiln between 300°C and 400°C for 4 hours. The mass lost was reported at 50-60% from which 20-30% was attributed to the original water lost as steam and we assume that the remaining loss percentage was the sum of water produced by the sulfation reactions, the initial water content of the sulfuric acid, and finally the evolved fumes of SO_2/SO_3.

The sulfated calcine is dissolved into hot water at 90°C with the lowest water-to-concentrate mass ratio [H_2O-to-S] of 2,200 kg per tonne. Actually, in order to ensure the maximum concentration of iron in solution, all the iron must remain in the ferric state as the maximum solubility of iron (III) sulfate in pure water can reach as high as 60 wt.% $Fe_2(SO_4)_3$ at room temperature (i.e., 300 g/L Fe(III)) [149]. The recovery yield for niobium was reported at 70-85 percent.

However, it was observed that performing the water dissolution under selected reducing condition enough to solubilize the manganese as Mn(II) but not too harsh to prevent the reduction of Fe(III) improved the niobium recovery by 10% according to the inventors. The selective dissolution is accomplished either by sparging sulfur dioxide gas or by adding scrap metallic iron.

The resulting pregnant leach solution containing the niobium values along with iron (III), titanium (IV), niobium (V), cerium (III), and uranium (VI) is intended to be processed further chemically by either selective precipitation or solvent extraction for monetizing all the metal values. Considering that all the rare earth industry relies on extracting the lanthanides from either nitrate or chloride based solutions the sulfate based media may pose certain challenges especially regarding separation by solvent extraction.

A prefeasibility study and a feasibility study were performed in 2014 and a 1:200-scale demonstration pilot plant was constructed on-site in Gabon by *Hatch Ltd.* (Mississauga, ON) to confirm the chemistry and to optimize the flow sheet [150].

Later it was announced in 2016 that the project has been idle and the company was looking for potential partner as the future commercial plant envisioned was supposed to beneficiate 2 million tonnes of run-of-mine (RoM) annually (RoM) to yield approximately 1 million tonnes of ore concentrate per annum.

10.7 VanadiumCorp-Electrochem Process

Presently, the main source of vanadiferous feedstocks for preparing vanadium pentoxide consists essentially of vanadiferous titano-magnetite (VTM)

[149] HODGMAN, C.D. (1962-1963) *Handbook of Chemistry and Physics, 44th Edition.* Chemical Rubber Publishing Co., Cleveland, OH, page 2038.

[150] NAZARI, G.; LAMOTTE, J.; RIES, M., AGIN, J.; TIZON, E.; KASHANI-NEJAD, S.; BELLINO, M.; and KRYSA, B. (2018) *Development of a Metallurgical Process for Eramet's Mabounié Nb-REE Project.* The Minerals, Metals and Materials Society (TMS) DAVIS, B.(ed.)(2018) *Extraction 2018*, August 26th-29th, 2018, Ottawa, ON, pp. 2353-2366.

occurring in magmatic layered gabbroic complexes such as the Bushveld complex found in the Republic of South-Africa, in Brazil and Australia; the yet untapped Lac Doré complex in the Chibougamau district (Quebec, Canada) and the Iron-T vanadium project in the Matagami area (Quebec, Canada). These ores depending on their geology contain usually between 0.4 wt. % and 1.5 wt.% V_2O_5. Additionally, some production originates from the processing of uranium ores such as carnotite from weathered sandstone deposits; the processing of phosphate ores; the petrochemical processing of bottom crude oils from Venezuela and Mexico; and the processing of vanadium-rich by-products from tar sands.

Presently, most of the vanadium pentoxide (V_2O_5) produced worldwide is derived from the mining of vanadiferous titano-magnetite (VTM) concentrates with vanadium values being locked as V(III) in the spinel phase coulsonite (FeV_2O_4), which forms a solid solution with ülvospinel (Fe_2TiO_4), and also in hematite phase (α-Fe_2O_3). The first step consists in the open pit mining from which the run-of-mine (RoM) is crushed, ground, screened and finally beneficiated by wet low intensity magnetic separation (WLIMS) in order to remove both silicates (e.g., Ca-plagioclase feldspars, pyroxenes) and sulfidic gangue minerals. The resulting VTM concentrate comprises an average chemical composition including 56-65 wt. % total iron (Fe), 8-12 wt. % TiO_2, and 0.8-2.0 wt. % V_2O_5. The resulting VTM is then typically processed by conventional pyrometallurgical and hydrometallurgical routes namely: (1) the soda ash process or salt roasting process ("Vametco Process"); and (2) the combination of smelting and slagging followed by soda ash roasting ("Highveld Process").

In the "Vametco process", used in South Africa, the vanadiferous titano-magnetite concentrate is mixed with sodium sulfate (Na_2SO_4) and sodium carbonate (Na_2CO_3). The charge was then roasted inside a rotary kiln at about 850°C. The hot clinker, discharged from the kiln, is quenched in cold water to produce a pregnant solution while the water leached calcine, containing most of the original iron, is disposed-off and landfilled. After the removal of sodium silicate using magnesium and aluminum sulfates (desilication), vanadium is precipitated as ammonium polyvanadate [$(NH_4)_2V_6O_{16}$] (APV) by adding ammonium sulfate at acidic pH or ammonium metavanadate (NH_4VO_3) (AMV) in accordance with the following reactions:

$$(NH_4)_2SO_4 + 6NaVO_3 + 2H_2SO_4 = (NH_4)_2V_6O_{16} + 2H_2O + 3Na_2SO_4$$

$$(NH_4)_2SO_4 + 2NaVO_3 = 2NH_4VO_3 + Na_2SO_4$$

After filtration, the precipitate is either calcined to produce red calcined vanadium pentoxide or further melted at temperatures above 690°C. followed by cooling in a drum flaker to yield black flakes with a metallic luster of fused vanadium pentoxide V_2O_5. Both products have a purity greater than 99.5 wt. % V_2O_5. The spent liquid effluents by-produced during these operations are evaporated in large sealed dams and the ammonium and sodium sulfate values are recycled back to the process.

Contrary to the "Vametco process", the "Highveld Process" already described in Section 10.3 consists in the smelting/slagging of the VTM followed by soda ash roasting.

This steel making process is energy intensive as the specific energy consumption for producing a slab of steel including smelting, refining and hot rolling is on average 6.38 MWh per tonne of steel slab. Moreover, the associated greenhouse gas emissions (GHGs) are on average 1.8 tonnes of carbon dioxide [151].

The specific energy consumption for producing vanadium pentoxide from pig iron is estimated as being about 3.0 MWh per tonne of vanadium pentoxide, including the crushing, grinding, and one or more of the following: roasting-leaching, roasting in a rotary-kiln, electric smelting, shaking-ladle, and basic oxygen furnace burning [152]

Despite the Vametco Process and the Highveld Process having gained worldwide industrial acceptance, especially the Highveld process because all the iron values are reported and recovered as high quality steel suitable for the most demanding industrial applications such as the production of High Strength Low Alloying Steels (HSLA), their main drawbacks are the following: (1) the large consumption of carbonaceous reductants that yield significant volumes of greenhouse gas (GHG) emissions during both pre-reduction, iron smelting/slagging, and steel making processes; (2) the significant consumption of expensive reductants such as anthracite coal and graphite electrodes during the arc smelting process; and (3) the high capital expenditures related to the building of the pre-reduction, arc smelting and roasting plants.

Beside the Highveld process that addresses the recovery of iron values as pig iron, the extraction of vanadium neglects the recovery of iron, titanium and silica during soda ash roasting.

Millions of tonnes of calcine stockpiles are by-produced from soda ash roasting are abandoned without any attempt to further processing them. The conventional process requires the sourcing of chemicals and raw materials (coal, electrodes, soda ash, and ammonia) overseas that impact the production cost significantly. The conventional process has significant water and energy consumptions. The conventional process exhibits a large carbon foot print due to the combustion taking place inside the rotary kiln using natural gas or pulverized coal as fuel. The Highveld-Evraz approach requires building an integrated smelter requiring major capital investment. Questionable profitability leads to the closure of the largest facilities worldwide (e.g., South Africa).

[151] FRUEHAN, R.J.; FORTONI, O.; PAXTON, H. W.; and BRINDLE, R. (2000) *Theoretical Minimum Energies to Produce Steel for Selected Conditions*. Carnegie Mellon University, Pittsburgh, PA.

[152] BLEIWAS, D.I. (2011) *Estimates of electricity requirements for the recovery of mineral commodities, with examples applied to sub-Saharan Africa*. U.S. Geological Survey (USGS), Open-file report 2011-1253, Reston, VA.

To address these important issues, and in order to monetize all the metal values, a novel chemical process was invented by the author, developed and patented jointly by *Electrochem Technologies & Materials Inc.* (Montreal, Canada) and *VanadiumCorp Resource Inc.* (Vancouver, Canada) [153].

The novel technology called the "VanadiumCorp-Electrochem Process Technology (VEPT)"[154] addresses the recovery of vanadium, iron, titanium, and silica values from a plethora of vanadiferous feedstocks such as vanadiferous titano-magnetite, iron ores, and concentrates such as magnetite and hematite, vanadium containing industrial wastes, and other industrial by-products also containing vanadium and iron.

The process is not yet industrial but it was tested successfully at the prototype and semi-pilot scale. A schematic flow diagram is depicted in Figure 35.

The VEPT process consists first to digest the finely ground 90% passing 45 micrometers (-325 mesh) vanadiferous titano-magnetite into concentrated sulfuric acid inside a brick-lined steel shell digester. Afterwards, the titano-magnetite concentrate is mixed with concentrated "winter" sulfuric acid (93 wt.% H_2SO_4).

The stoichiometric (theoretical) sulfuric acid-to-solid mass ratio [H_2SO_4-to-S] of 1,584 kg (100 wt.% H_2SO_4) per tonne of dry concentrate is used. This corresponds to a practical (actual) sulfuric acid-to-solid ratio [A-to-S] of 1,700 kg (93 wt.% H_2SO_4) per tonne of dried VTM.

The necessary heat required to warm the charge is simply obtained by injecting water directly into the mixture until reaching a final concentration of acid close to 90 wt.% H_2SO_4. The significant heat released by the enthalpy of hydration of the sulfuric acid, brings within few minutes the overall temperature inside the charge to 80-90°C, and then it triggers the exothermic sulfation reaction with the temperature of the charge reaching 180-210°C in a matter of minutes and then lasting for several hours due to the good thermal insulation provided by the refractory and acid proof bricks.

The key figures related to the VEPT process are summarized and reported in Table 66 hereafter.

---

[153] CARDARELLI, F. (2018) *Metallurgical and chemical processes for recovering vanadium and iron values from vanadiferous titanomagnetite and vanadiferous feedstocks*. PCT International Patent Application WO 2018/152628 (A1) (Electrochem Technologies & Materials Inc. and VanadiumCorp Resource Inc.), August 30th, 2018.

[154] CARDARELLI, F. (2021) *Metallurgical and chemical processes for recovering vanadium and iron values from vanadiferous titanomagnetite and vanadiferous feedstocks*. U.S. Patent 10,947,630 B2 (VanadiumCorp Resource Inc.), March 16th, 2021.

Table 66 – Sulfation of vanadiferous titano-magnetite: key figures

TITANO-MAGNETITE (concentrate)			
Chemical composition		Strength H_2SO_4 (initial)	**93%**
Fe_2O_3	51.78 wt.%	Strength H_2SO_4 (attack)	90%
FeO	27.46 wt.%	[H_2SO_4-to-Solid](Theor.)	**1,584** kg/tonne
TiO_2	11.40 wt.%	Mass excess of acid	**0%**
SiO_2	3.10 wt.%	[H_2SO_4-to-Solid](Actual)	**1,584** kg/tonne
Al_2O_3	2.63 wt.%	[H_2SO_4-to-TiO_2](Actual)	**13,892** kg/tonne
V_2O_5	1.13 wt.%	[Acid-to-Solid](Actual)	**1,703** kg/tonne
CaO	0.55 wt.%	Solids pulp density	37%
MgO	0.82 wt.%	c_p (solid)	673 J/kg/K
Cr_2O_3	0.08 wt.%	c_p (acid)	1467 J/kg/K
	98.87 wt.%	c_p (pulp)	**1193** J/kg/K
		Δh Sulfation	**-1004** kJ/kg of solid
Mass of solids (dry)	**1000** kg	Δh Hydration	-62 kJ/kg of acid
Mass of sulfuric acid	1703 kg	Δh Vaporization	2256 kJ/kg of steam
Mass of water (injected)	57 kg	Δh Total	**-258** kJ/kg of mixture
Mass of pulp	**2760** kg	Adiabatic temp rise ΔT	**216 K EXOTHERMIC**

This peculiar behavior allows operating quasi-autogenously until the baking is complete and takes usually 3 to 4 hours with the production of a hard and porous sulfation cake. The dissolution of the sulfation cake is performed around 75°C with slightly acidified water to ensure sufficient free acidity remains in solution to prevent the hydrolysis of Fe(III), V(IV) and Ti(IV) with a water-to-solid mass ratio [H_2O-to-S] is close to 2,000 kg of water per tonne of sulfation cake to ensure the proper and fast dissolution of all the water soluble metal sulfates.

After separating the insoluble solids, mostly coarser unreacted magnetite, silica, and silicates, by gravity settling and filter pressing this yields a concentrated pregnant solution. Because most of the iron is in the trivalent state, the pregnant solution is reduced electrochemically under high cathode current density using a proprietary divided electrolyzer manufactured by *Electrochem Technologies & Materials Inc.* (Montreal Canada) with a plate and frame configuration. The reduction in performed until some titanium is reduced to Ti(III) to ensure the iron in the reduced pregnant solution is fully reduced and is protected towards air oxidation.

The reduced liquor is then subjected to the chilling down to 5°C and crystallization to yield crystals of copperas (ferrous sulfate heptahydrate). After removing the copperas, the process further comprises removing titanium from the iron depleted solution by hydrolysis using the Blumenfeld method thereby producing a vanadium-bearing pregnant solution.

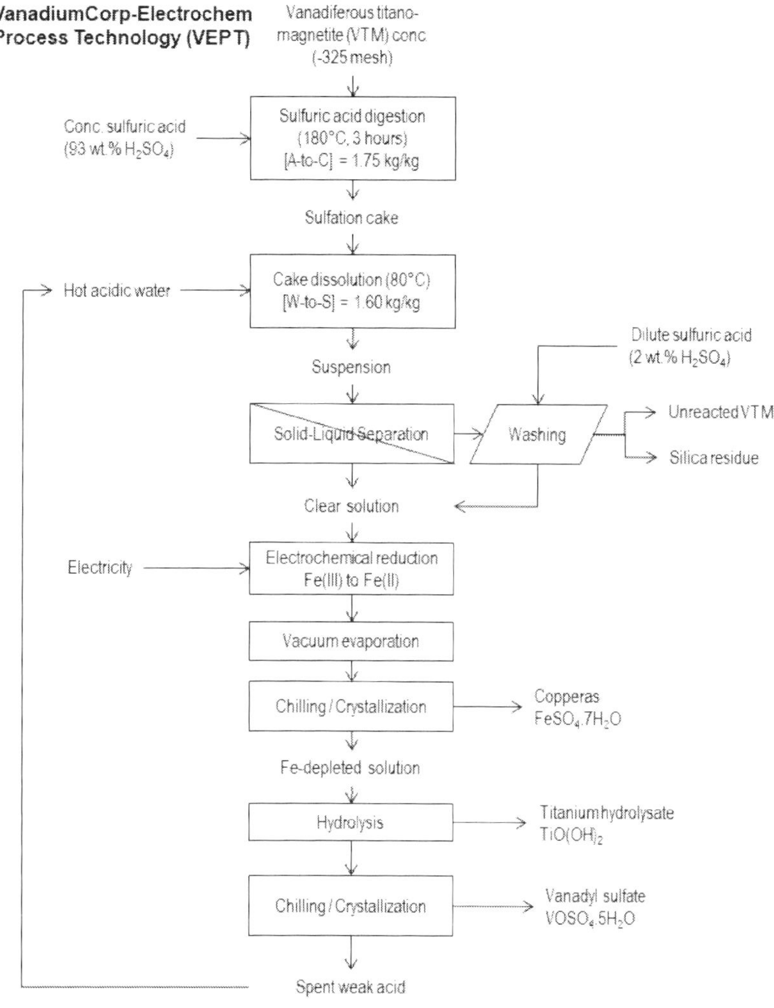

Figure 35 – Flow diagram for the VanadiumCorp-Electrochem Process Technology

The further concentration by evaporation and a sequence of chilling and crystallization steps yields vanadyle sulfate pentahydrate to be purified or to be converted eventually into various vanadium chemicals (AMV, APV, V_2O_5) that could be used as precursors for the preparation of vanadium electrolyte used for vanadium redox flow batteries (VRFB).

For the sake of clarity, it is important to mention that not all the vanadium feedstocks are equivalent and some are not necessarily suitable to be processed by the VEPT. Actually, the VEPT addresses and targets mainly iron-rich and low grade vanadium vanadiferous materials while vanadium-rich raw materials containing low concentrations of iron and titanium such as spent catalysts, spent residues from oil refineries or containing more than 30 wt.% V_2O_5 such as secondary vanadium-slags are usually better processed using the conventional alkaline roasting or by the salt roasting routes.

Actually, the major limitations for using the VEPT are certain raw materials containing a high carbon content either as elemental carbon (C) or aliphatic or aromatic hydrocarbons (C_xH_y), an elevate free alkalinity (Na_2O), or a strong basicity (CaO, MgO). The first requires the removal of carbon-based materials by solvent extraction step or a direct combustion treatment that poses certain environmental issues while the two latter consume irretrievably sulfuric acid values unless there is a particular economical interest locally for selling the sodium sulfate dodecahydrate (Glauber's salt), the gypsum, and the magnesium sulfate heptahydrate, that are by-produced (see Section 11).

In these particular cases, the high reactivity of strongly alkaline or basic raw materials requires to use an elevate actual sulfuric acid-to-solid mass ratio [H_2SO_4-to-S] up to 20,000 kg (100 wt.% H_2SO_4) per tonne in some instance to absorb the important amount of heat released hence preventing a thermal runaway.

11 Sulfation By-products, Effluents, and Wastes

Because most of the raw materials, ores and concentrates processed using sulfation methods always contain significant concentrations of ubiquitous elements such as iron, calcium, and sodium inevitably the by-product's and effluents end up as sulfates of iron, sodium and calcium that needs

11.1 Iron sulfate(s)

Because the ubiquitous presence of iron oxides in most ores and concentrates for all commodities, it becomes obvious that the sulfation of such raw materials and feedstocks yields significant amount of solid wastes and liquid effluents containing ferrous and/or ferric sulfates. The ratio of iron (III) and iron (II) strongly depends on the speciation of iron in the raw material, the redox conditions existing during the process, the storage, the presence of reducing species such as hydrogen sulfide, sulfur dioxide or even elemental sulfur during the sulfuric acid digestion and sulfation baking or the intentional reduction of the ferric iron by adding iron metal scrap, injecting sulfur dioxide or other scarcer chemicals such as barium sulfide ("black ash") or oxalic acid in some instances the latter occurring in bauxite residues.

Therefore, the handling of such iron-rich wastes and effluents poses serious issues to the chemical and metallurgical industries. Several technical solutions have been used while some novel approaches needs still to be validated at a large scale.

11.1.1 Neutralization and Disposal

The simplest option is to dispose and landfill often after proper neutralization with lime and limestones the effluents yielding large tonnages of sludge made of tainted gypsum. Unfortunately, due the instability towards weather usually bring a long-term liability associated with iron residue disposal not to mention that due to catastrophic consequences with the collapse of dam, the poor environmental acceptability of such approach, and the more and more stringent environmental regulations enforced worldwide such option is the least desired.

The neutralization of spent sulfuric acid, waste acid or weak acids origination from the chemical and metallurgical industries are usually performed using either lime or limestone [155].

The mass of 100% neutralizing agent (e.g., dolime, lime, slacked lime, caustic soda, caustic potash, potash, and soda ash) required per unit mass of 100% sulfuric acid, that is, the mass of diluted acid times the mass percentage of H_2SO_4, can be determined quickly using the practical graph depicted in Figure 36.

From the graph, we can see clearly that dead burned magnesia is the most efficient neutralizing agent as theoretically only 41 kg of MgO are consumed per 100 kg of H_2SO_4 followed by dolime and quicklime with consumptions of 50 kg and 57 kg per 100 kg of H_2SO_4 respectively.

[155] BOYTON, R.S. (1980) *Chemistry and Technology of Limestone.* John Wiley & Sons, Inc., New York, NY pp. 211-213.

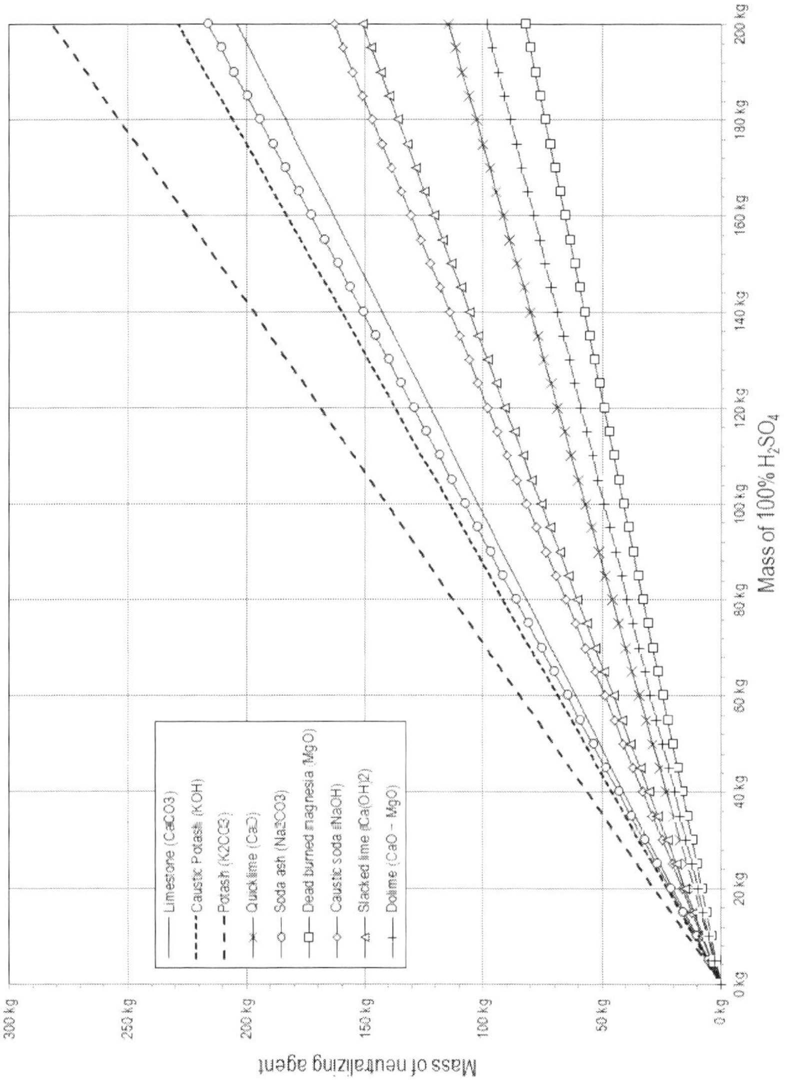

Figure 36 – Neutralization equivalents for 100% sulfuric acid

The removal of iron is usually performed by precipitation of iron as ferric hydroxide, $Fe(OH)_3$ because it exhibits an extremely low product of solubility. Moreover, the optimal pH for its precipitation of ferric hydroxide usually ranges between 2.2 and 4.0. Therefore removal of iron as ferric hydroxide avoids the precipitation of other valuable metal hydroxides. The reaction for the precipitation along with its product of solubility is given below:

$$Fe^{3+} + 3OH^- \rightarrow Fe(OH)_3\downarrow \text{ with } Ks = (Fe^{3+})(OH^-)^3 = 2.79 \times 10^{-39}$$

For an efficient precipitation, it is important however, that complexing agents that prevent the precipitation of ferric hydroxide must not be present in the liquid effluent. The most common industrial complexing agents includes phosphoric acid, oxalic acid, and hydrofluoric acid that all form strong ligands with the ferric cations [156] thus preventing or hindering the precipitation.

Moreover, reducing agents must be avoided at all costs because, by reducing ferric back to ferrous cations, they prevent the precipitation of ferric hydroxide by forming ferrous hydroxide that starts to precipitate only at pH above 5.5. Therefore, the removal of iron must always be performed after all the iron is oxidized into its ferric state by forced aeration, injection of oxygen gas, or eventually using hydrogen peroxide.

Usually, the removal of iron as ferric hydroxide is very efficient with less than 10 ppm wt. (i.e., 10 mg/kg of solution) of residual iron in the liquor at the end. Nevertheless, the technique is time consuming because of the colloidal nature of the gelatinous precipitate that can clog the cloth of the filter presses.

Despite in all cases, all the iron ends up as precipitated ferric hydroxide; the following salts are also produced co-currently in large amount such as gypsum, and sodium sulfate that are discussed in detail hereafter.

11.1.2 Precipitation of Natro-jarosite (Jarofix)

An alternative to the simple precipitation of ferric hydroxide is the selective precipitation of natro-jarosite with chemical formula $NaFe_3(SO_4)_2(OH)_6$. By contrast with ferric hydroxide, the precipitate of natro-jarosite brings additional benefits compare to the previous alkalization as the iron-bearing precipitate is relatively stable, exhibits a much larger particle size thus it is easy to filter and the amount of reducing agent is largely minimized not to mention that some sulfuric acid is recycled.

The advantages of the jarosite process include the production of readily filterable iron precipitates that would make it easier to separate by cross flow separation methods along with minimum losses of other metals entrapped in the jarosite precipitate; simultaneous control of sulfate and alkali metal cations; and the ease of integration in an existing hydrometallurgical process.

[156] RINGBOM, A. (1963) *Complexation in Analytical Chemistry.* John Wiley & Sons, Inc., New York, NY, pp. 272-350.

In its simplest form, the jarosite process involves the addition of an alkali source usually sodium carbonate or soda ash (Na_2CO_3) or sodium sulfate (Na_2SO_4), seeds of jarosite that accelerate the rate of jarosite precipitation, and a neutralizing agent to a ferric iron-rich, and hot acid solution.

The neutralizing agent is required to control both the residual acidity from the hot acid solution and the free acid produced by the hydrolysis of the ferric cations according to the following chemical reaction:

$$3Fe_2(SO_4)_3 + Na_2SO_4 + 12H_2O \rightarrow 2NaFe_3(SO_4)_2(OH)_6\downarrow + 6H_2SO_4$$

In practice, the precipitation is performed at a low pH of 1.7 and at a temperature of 98°C. It is important to note, that the practical concentration cut-off at the end of the precipitation is on average 5g/L of Fe(III) but a longer residence time or different operating conditions for instance using NH_3 instead of NaOH, could eventually reduce that threshold concentration well below that level.

For instance, the Jarofix® process performed by *Canadian Electrolytic Zinc Ltd.* or *CE-Zinc* (Valleyfield, QC) the jarosite mineral phase precipitates during the leaching of zinc ferrites ($ZnFe_3O_4$) and it is mixed with Portland cement, lime, and water. The chemical reaction generates a chemically and physically stable material, reducing the amount of neutralizing agent while offering concomitant processing advantages. Since 1998, it became a benchmark for the zinc industry with 160,000 tonnes produced per annum solely in Quebec [157].

11.1.3 Copperas for Wastewater Treatment

The copperas by-produced in the titanium dioxide pigment industry and the iron and steel making industries, is usually valuable enough to be recovered from the iron-rich sulfate solutions and liquors. For instance, in the titanium pigment industry, every plant using the sulfate route that is used to process beach sand ilmenite with a minimum of 56 wt.% TiO_2 generates large volumes of iron sulfate that needs to be addressed from an environmental and economic standpoint.

In practice, after reduction of the ferric iron by adding scrap metallic iron, the ferrous sulfate solution is clarified by settling and cross-flow filtration prior to cooling in order to remove insoluble residues. The iron-rich solution is then cooled down using an ammonia refrigeration unit and sent to vertical crystallizers to allow for the crystallization of iron (II) sulfate heptahydrate with chemical formula, $FeSO_4.7H_2O$, CAS No. [7782-63-0], also called **copperas** or **green vitriol** which is later removed from the crystal slurry by means of disk bowl centrifuges.

---

[157] CHEN, T.T.; and DUTRIZAC, J.(2001) Jarofix: addressing iron disposal in the zinc industry. *JOM: the journal of the Minerals, Metals & Materials Society*, **53**(12):32-35

The estimated tonnage of copperas produced annually in titanium pigment industry is approaching 8.65 million tonnes while if other industries are included the total equivalent of copperas produced annually is close to 22 million tonnes with the breakdown by geographical location and industries reported in Table 67.

Table 67 – World sources for copperas (tonnes per year)

Geographical region	Country	Copperas from titanium pigment industries	Spent pickling liquors from iron and steel industries	Pregnant solutions from the of nonferrous metals	Total per region
North America	USA	none	500,000	1,000,000	**2,580,000**
	Canada	20,000	60,000	1,000,000	
South America	Brazil	20,000	150,000	1,000,000	**1,200,000**
	Argentina		25,000		
	Others		25,000		
Europe	Finland	600,000	1,200,000	500,000	**3,380,000**
	Germany	400,000			
	Spain	230,000			
	Others	450,000			
Australasia	Australia	none	40,000	1,500,000	**1,540,000**
	New Zealand				
	Others				
Asia	China	6,400,000	3,000,000	1,000,000	**11,950,000**
	Japan	250,000	500,000	1,000,000	
	Others	300,000	500,000		
Total worldwide by industries =		**8,670,000**	**6,000,000**	**7,000,000**	**21,670,000**

The copperas being either stockpiled on-site as it was the case in Finland, near the former *Kemira Oy* plant or sold and used as flocculating agent for potable and wastewater treatment, and as reducing chemical to address the mitigation of hexavalent chromium in the cement industry.

The copperas product is then sold as either as: (1) flocculating agent used either directly or after conversion of an iron (III) product for the treatment of wastewaters, (2) fertilizer in agriculture for prevention of iron chlorosis in plants grown on iron deficient soils or for moss control [158, 159], and finally (3) as an additive in the cement manufacturing process.

Actually in the latter, ferrous sulfate or moist copperas is used as reducer of hexavalent chromium (CrO_4^{2-}) which is always present in traces in cement into insoluble trivalent chromium hydroxide. This is achieved in order to eliminate the hexavalent chromium that causes serious occupational issues and health hazards thereby preventing the risk of chrome dermatitis for exposed workers and end users. The use of iron (II) sulfate as additive in cement for chromate reduction is described in these references [160, 161].

The estimated amount of copperas sold annually as flocculating agent is 2 million tonnes, while 330,000 tonnes are sold as fertilizer in agriculture, and only 70,000 tonnes are sold as additive in cement to stabilize the hexavalent chromium.

Therefore, the total amount of ferrous sulfate heptahydrate used in these markets is less than 3 million tonnes that must be compared to the 21.65 million tonnes of copperas equivalent generated worldwide annually.

11.1.4   Crystallization of Copperas

The crystallization of ferrous sulfate heptahydrate (copperas) is an efficient way to recover the iron values from ferrous sulfate rich liquors once all the ferric iron is reduced back to ferrous iron. Considering that crystallizing copperas not only carries also the removal of water but provide a clean saleable product suitable to be converted to red iron pigment or metallic iron from processes described previously.

11.1.4.1   Crystallization by Cooling

Let's consider the recovery of crystals of ferrous sulfate heptahydrate with the chemical formula $FeSO_4.7H_2O$ by cooling a mass, M_0, in kg (lb.) of a hot concentrated mother liquor at an initial temperature T_1 in °C(°F) and containing a mass percentage, w_0, expressed as the anhydrous ferrous sulfate ($FeSO_4$). The initial masses of anhydrous ferrous sulfate and water contained in the mother liquor are simply calculated as follows:

$$m_{FeSO_4}^{0} = w_0 M_0 \quad \text{and} \quad m_{H2O}^{0} = (1 - w_0)M_0$$

---

[158] KOENIG, R. and KUHNS, M. (1996) *Control of Iron Chlorosis in Ornamental and Crop Plants*. Utah State University, Salt Lake City, UT, August 1996.

[159] HANDRECK, K. (2002) *Gardening Down Under: A Guide to Healthier Soils and Plants, Second Edition*. CSIRO Publishing, Collingwood, Victoria, Australia, pp. 146–47.

[160] MANNS, W.; LASKOWSKI, Ch. (1999) Eisen(II)sulfat als Zusatz zur Chromatreduzierung" BE-Z: *Beton Journal*, Volume 2 (1999) pp. 78-85.

[161] KEHRMANN, A (2008) *Hydraulic Binder and a Chromate Reducer and Use Thereof*. U.S. Patent Application 2008/0282939.

---

with the overall mass balance:

$$M_0 = m_{FeSO_4}{}^0 + m_{H2O}{}^0$$

The exact hydrate composition with definite number of molecules of water can be determined from the examination of the solid-liquid phase diagram of the binary system: $FeSO_4$-H_2O depicted in Figure 37 with several that can be crystallized depending on the final cooling temperature.

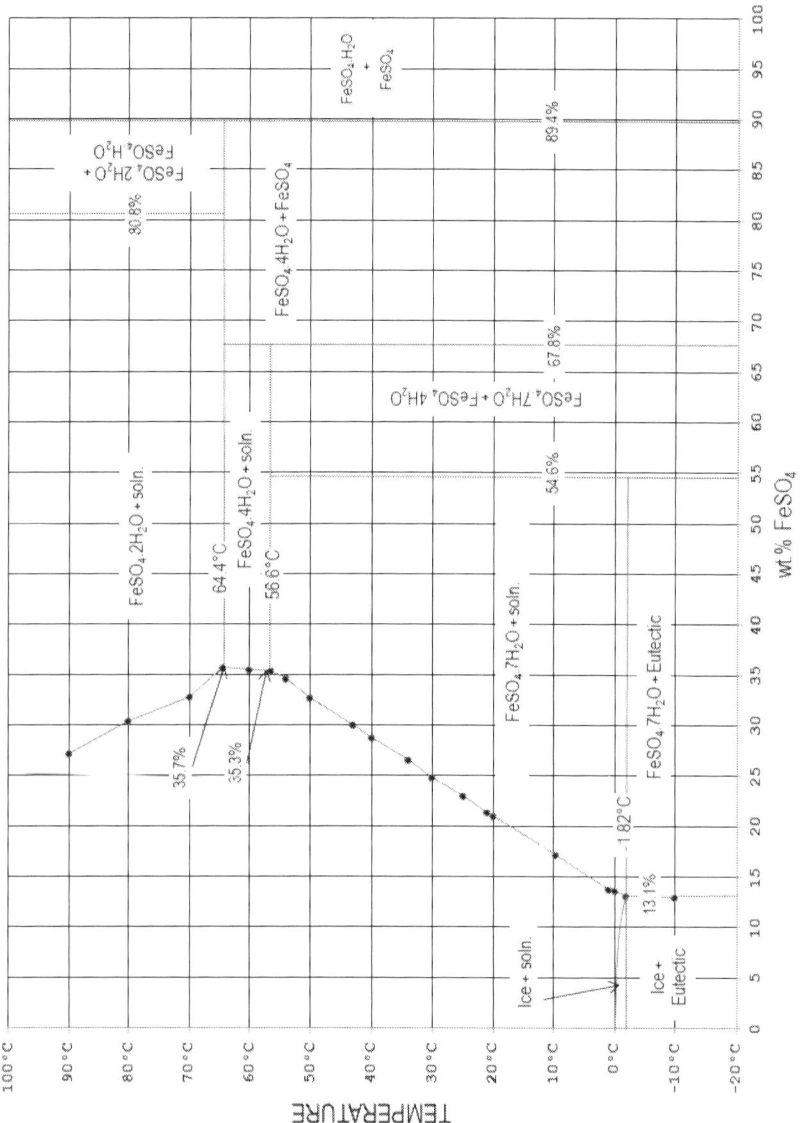

Figure 37 – Phase diagram of the binary FeSO$_4$-H$_2$O system

Therefore upon cooling at a final temperature T_2 expressed in °C(°F), the crystallization yields a mass m_C of crystals with mass percentage, w_C, of anhydrous ferrous sulfate that are in thermal equilibrium with a mass m_L of depleted solution having a mass percentage w_L, of FeSO$_4$.

The overall mass balance imposes: $M_0 = m_C + m_L$

While the individual mass balances for each components are given by:

For crystals of ferrous sulfate hydrate: $\qquad w_0 M_0 = w_C m_C + w_L m_L$

For the water: $\qquad (1 - w_0)M_0 = (1 - w_C)m_C + (1 - w_L)m_L$

Therefore combining and rearranging the above equations, it is possible to calculate the mass fractions of liquid (i.e., depleted solution) (φ_L), the mass fraction of solids (i.e., crystals of a given ferrous sulfate hydrate) (φ_C) and the ratio of the mass of solution over the mass of crystals at the final temperature T_2:

$$\varphi_L = m_L/M_0 = (w_0 - w_C)/(w_L - w_C)$$

$$\varphi_C = m_C/M_0 = (w_L - w_0)/(w_L - w_C)$$

$$R_{LC} = m_L/m_C = (\varphi_L/\varphi_C) = (w_0 - w_C)/(w_L - w_0)$$

The above equations are essentially the mathematical expressions of the *inverse lever-arm rule* used for the determination of mass fractions graphically using binary phase diagrams.

Therefore, the final masses of depleted solution and crystals that produced by cooling are simply obtained:

$$m_C = \varphi_C M_0 = [(w_L - w_0)/(w_L - w_C)]M_0$$

$$m_L = \varphi_L M_0 = [(w_0 - w_C)/(w_L - w_C)] M_0$$

The dimensionless *recovery yield* for the anhydrous ferrous sulfate, denoted R_{FeSO4}, that is, the mass of $FeSO_4$ contained in the hydrated crystals vs. the initial amount of anhydrous $FeSO_4$ in the mother liquor is given by the equation:

$$R_{FeSO4}(\%) = 100[m_{FeSO4}(C)/m_{FeSO4}^0] = 100(w_C/w_0)[(w_L - w_0)/(w_L - w_C)]$$

On the other hand, the **water retention in the depleted solution** is given by:

$$R_{H2O}(\%) = 100[m_{H2O}(L)/m_{H2O}^0] = 100(w_C/w_0)\{(1 - w_L)[(w_0 - w_C)/(w_L - w_C)]\}$$

All the previous equations are summarized in Table 68 hereafter.

Table 68 – Equations for the crystallization of copperas by cooling

Initial (at T_1)	Crystals (IN)		Mother liquor (FEED)	
	Initial mass of solids: none		Mass percentage of anhydrous salt: w_0 Mass of mother liquor: M_0	
	Anhydrous salt (FeSO$_4$)	Water (H$_2$O)	Anhydrous salt (FeSO$_4$)	Water (H$_2$O)
	None	None	$m_{FeSO4}{}^0 = w_0 M_0$	$m_{H2O}{}^0 = (1 - w_0)M_0$
Final (at T_2)	Crystals (OUT) Mass percentage of anhydrous salt: w_C Mass of crystals: $m_C = \varphi_C\, M_0 = [(w_L - w_0)/(w_L - w_C)]M_0$		Depleted solution (OUT) Mass percentage of anhydrous salt: w_L Mass of solution: $m_L = \varphi_L\, M_0 = [(w_0 - w_C)/(w_L - w_C)]\, M_0$	
	Anhydrous salt (FeSO$_4$)	Water (H$_2$O)	Anhydrous salt (FeSO$_4$)	Water (H$_2$O)
	$m_{FeSO4}(C) = w_C[(w_L - w_0)/(w_L - w_C)]\,M_0$	$m_{H2O}(C) = (1 - w_C)[(w_L - w_0)/(w_L - w_C)]\, M_0$	$m_{FeSO4}(L) = w_L[(w_0 - w_C)/(w_L - w_C)]\, M_0$	$m_{H2O}(L) = (1 - w_L)[(w_0 - w_C)/(w_L - w_C)]\, M_0$

Recovery yield for anhydrous salt:

$$R_{FeSO4}\,(\%) = 100\; m_{FeSO4}(C)/m_{FeSO4}{}^0 = 100(w_C/w_0)[(w_L - w_0)/(w_L - w_C)]$$

Example: A steel pickling plant generates a mass flow rate of 10,000 kg per day of a warm and saturated aqueous solution of ferrous sulfate at 54°C (129.2°F) which is cooled down to 10°C (50°F) in order to crystalize out crystals of ferrous sulfate heptahydrate FeSO$_4$.7H$_2$O (copperas) to be sold to a nearby waste water treatment plant. What are the mass flow rates of crystals of copperas and that of depleted solution produced daily? Finally, what is the ferrous sulfate recovery yield and water retention in the final depleted solution?

First, from the phase diagram of the binary system FeSO$_4$-H$_2$O, it is possible to retrieve the mass percentages of anhydrous ferrous sulfate for the saturated solution at 54°C and that for the cold depleted solution at 10°C, that is, 34.52 wt.% and 17.11 wt.% FeSO$_4$ respectively. Secondly, applying the inverse lever-arm rule or the previous equations, the mass fractions of liquid and solid, that are in equilibrium at 10°C, are: $\varphi_L = (54.60 - 34.52)/(54.60 - 17.11) = 53.56$ wt.% and $\varphi_C = (34.52 - 17.11)/(54.60 - 17.11) = 46.44$ wt.% respectively. Thirdly, these mass ratios allow to calculate the mass flow rate of copperas crystals: 4,644 kg/day which contains 54.6 wt.% FeSO$_4$ and that of the cold depleted solution: 5,356 kg/day containing only 17.11 wt.% FeSO$_4$. The recovery yield of ferrous sulfate, R_{FeSO4} is 73.5% while the retention of water R_{water} in the cold solution is only 67.8% because the crystallization of copperas removes seven moles of water per mole of anhydrous salt. These calculations are all summarized in Table 69.

This example shows the benefit of recycling spent pickling liquors that are usually saturated with respect to ferrous sulfate by simply cooling them in order to remove ca. 27% of its iron content and then rejuvenating them by simply adding fresh concentrated sulfuric acid.

Table 69 – Crystallization of copperas by cooling

Temperature	MOTHER LIQUOR COMPOSITION						CRYSTAL COMPOSITION			CRYSTALS		LIQUOR (FEED)	
	$FeSO_4$	H_2O (total)	Copperas	H_2O (free)	Mass $FeSO_4$ per 100 kg H_2O	Mass copperas per 100 kg H_2O	$FeSO_4$	H_2O	IN	NONE		10,000 kg/day	
										$FeSO_4$	H_2O	$FeSO_4$	H_2O
54.0°C 129.2°F	34.52%	55.48%	63.18%	36.82%	52.72	96.48	NONE	NONE		0 kg	0 kg	3,452 kg	6,548 kg
Temperature	DEPLETED SOLUTION COMPOSITION						CRYSTAL COMPOSITION		OUT	CRYSTALS		SOLUTION	
										$\varphi_C = 46.44\%$		$\varphi_L = 53.56\%$	
										4,644 kg/day		5,356 kg/day	
	$FeSO_4$	H_2O (total)	Copperas	H_2O (free)	Mass $FeSO_4$ per 100 kg H_2O	Mass copperas per 100 kg H_2O	$FeSO_4$	H_2O		$FeSO_4$	H_2O	$FeSO_4$	H_2O
10°C 50.0°F	17.11%	82.39%	31.31%	68.69%	20.64	37.6	54.60%	45.40%		2,536 kg	2,108 kg	916 kg	4,440 kg
										$R_{FeSO_4} = 73.5\%$		$R_{H_2O} = 67.8\%$	

11.1.4.2   Crystallization by Evaporative Cooling

Instead of simply cooling a brine or solution of a ferrous hydrated salt for performing the recovery of the salt, it is also possible to concentrate first the solution by removing a given mass of water as steam by evaporation either under atmospheric pressure (normal boiling) or under reduced pressure (vacuum) and then conducting the cooling afterwards. This approach is particularly suited for diluted solutions.

Using again as example an aqueous solution of ferrous sulfate, if a mass of water, m_{steam}, in kg (lb.) is removed as steam, the mass percentage of ferrous sulfate in the concentrated mother liquor increases and it can be expressed as follows:

$$m_{FeSO4} = w_0 M_0 / (M_0 - m_{steam})$$

Therefore, the calculations to obtain the mass of solution and the mass of crystals made of ferrous sulfate hydrate are simply obtained and are summarized in Table 70 hereafter.

Table 70 – Equations for crystallization of copperas by evaporative cooling

Initial (at T_1)	Crystals (IN)		Mother liquor (FEED)	
	Initial mass of solids: none		Mass percentage of anhydrous salt: w_0 Mass of mother liquor: M_0	
	Anhydrous salt (FeSO$_4$)	Water (H$_2$O)	Anhydrous salt (FeSO$_4$)	Water (H$_2$O)
	None	None	$m_{FeSO4}^0 = w_0 M_0$	$m_{H2O}^0 = (1 - w_0)M_0$
Evaporation (at T_2)	Steam (OUT)		Concentrated mother liquor	
	Mass of steam removed: m_{STEAM}		Mass percentage of anhydrous salt: w_{MX} Mass of conc. solution: $M = M_0 - m_{STEAM}$	
	Anhydrous salt (FeSO$_4$)	Water (H$_2$O)	Anhydrous salt (FeSO$_4$)	Water (H$_2$O)
	None	None	$w_{FeSO4} = w_0 M_0/(M_0 - m_{steam})$ $m_{FeSO4} = w_0(M_0 - m_{steam})$	$w_{H2O} = (1 - w_0)M_0/(M_0 - m_{STEAM})$ $m_{H2O} = m_{H2O}^0 - m_{STEAM}$
Final (at T_2)	Crystals (OUT)		Depleted solution (OUT)	
	Mass percentage of anhydrous salt: w_C Mass of crystals: $m_C = \varphi_C M = [(w_L - w_{FeSO4})/(w_L - w_C)]M$		Mass percentage of anhydrous salt: w_L Mass of solution: $m_L = \varphi_L M = [(w_{FeSO4} - w_C)/(w_L - w_C)]M$	
	Anhydrous salt (FeSO$_4$)	Water (H$_2$O)	Anhydrous salt (FeSO$_4$)	Water (H$_2$O)
	$M_{FeSO4}(C) = w_C[(w_L - w_{FeSO4})/(w_L - w_C)]M$	$m_{H2O}(C) = (1 - w_C)[(w_L - w_{FeSO4})/(w_L - w_C)]M$	$M_{MX}(L) = w_L[(w_{FeSO4} - w_C)/(w_L - w_C)]M$	$m_{H2O}(L) = (1 - w_L)[(w_{FeSO4} - w_C)/(w_L - w_C)]M$
Recovery yield for anhydrous salt: $R_{FeSO4}(\%) = 100\, m_{FeSO4}(C)/m_{FeSO4}^0 = 100(w_C/w_0)[(w_L - w_0)/(w_L - w_C)]$				

Example: A chemical plant produces 25,000 kg per hour of a warm solution containing 25 wt.% ferrous sulfate at 60°C which is evaporated under vacuum by removing 5,000 kg per hour of steam and the concentrated solution is then cooled down to 10°C in order to crystalize out crystals of ferrous sulfate heptahydrate (copperas). What are the mass flow rates of crystals and depleted solution obtained after evaporative cooling? What are the recovery yield and water retention?

First, removing 5,000 kg/h of water as steam from the mother liquor produces 20,000 kg per hour of concentrated liquor. The new concentration of ferrous sulfate in that liquor increases and it becomes: 25 (25,000/20,000) = 31.25 wt.%. From the phase diagram of the binary system FeSO$_4$-H$_2$O, the mass percentages of ferrous sulfate in the final depleted solution at 10°C is: 17.11 wt.% $FeSO_4$. Then applying the inverse lever arm rule, the fractions of liquid and solid in equilibrium after cooling are: φ_L = (54.6 - 31.25)/(54.6 − 17.11) = 61.98% and φ_C = (31.25 − 17.11)/(54.6 − 17.11) = 38.08%, then the two mass flow rates: 7,604 kg/day of copperas crystals and 12,396 kg/day of depleted solution. The recovery yield of ferrous sulfate is 66.1% while the retention of water in the solution is only 54.8% because most of the water was remove first by thermal evaporation. These results are summarized in Table 71 hereafter.

Table 71 – Crystallization of copperas by evaporative cooling

Temperature	MOTHER LIQUOR COMPOSITION						CRYSTAL COMPOSITION		IN	CRYSTALS		LIQUOR (FEED)	
										NONE		25,000 kg/day	
60.0°C 140°F	FeSO$_4$	H$_2$O (total)	Copperas	H$_2$O (free)	Mass FeSO$_4$ per 100 kg H$_2$O	Mass copperas per 100 kg H$_2$O	FeSO$_4$	H$_2$O		FeSO$_4$	H$_2$O	FeSO$_4$	H$_2$O
25.00%	75.00%	45.75%	54.25%	33.33	61.0	NONE	NONE		0 kg	0 kg	6,250 kg	18,750 kg
	CONCENTRATED MOTHER LIQUOR						CRYSTAL COMPOSITION		OUT	STEAM			
										5,000 kg/day			
EVAPORATION	FeSO$_4$	H$_2$O (total)	Copperas	H$_2$O (free)	Mass FeSO$_4$ per 100 kg H$_2$O	Mass copperas per 100 kg H$_2$O	FeSO$_4$	H$_2$O					
31.25%	68.75%	57.19%	42.81%	45.45	83.2	NONE	NONE					
Temperature	DEPLETED SOLUTION COMPOSITION						CRYSTAL COMPOSITION		OUT	CRYSTALS		SOLUTION	
										φ_C = 38.02%		φ_L = 61.98%	
										7,604 kg/day		12,396 kg/day	
10°C 50°F	FeSO$_4$	H$_2$O (total)	Copperas	H$_2$O (free)	Mass FeSO$_4$ per 100 kg H$_2$O	Mass copperas per 100 kg H$_2$O	FeSO$_4$	H$_2$O		FeSO$_4$	H$_2$O	FeSO$_4$	H$_2$O
17.11%	82.89%	31.31%	68.69%	20.64	37.8	54.30%	45.70%		4129 kg	3475 kg	2121 kg	10275 kg
										R_{FeSO_4} = 66.1%		R_{H_2O} = 54.8%	

11.1.5 Roasting and Decomposition of Iron Sulfates

Because, the lost values of sulfuric acid immobilized in the copperas crystals, several pigment plant producers performed the regeneration of sulfuric by roasting the copperas in air producing sulfur trioxide sent to a sulfuric acid plant and by-producing high quality red iron oxide pigment.

However, the regeneration of sulfuric acid from copperas or spent pickling liquors produces sulfuric acid at much higher costs, compared to the acid produced from conventional raw materials (sulfur, sulfide ores). But the crucial question of economics cannot be answered by production costs alone. With the enforcement of restrictive pollution control laws, costs for the ecologically acceptable disposal of spent acids must be added when comparisons are made with the costs for conventional sulfuric acid production.

Actually, in Europe, because of stringent environmental regulations, most of the European titanium dioxide pigment producers have to cope with the treatment of their own copperas, the same applied for the spent pickling liquors generated from iron and steel making operations. For instance, since the late 1980s in Germany, the titanium and steel industries are no longer permitted to dump the iron sulfate residues or spent pickling liquors into the sea. These regulations have forced the entire German industry to develop a new technology. The current technology is based entirely on an energy demanding thermal roasting process [162].

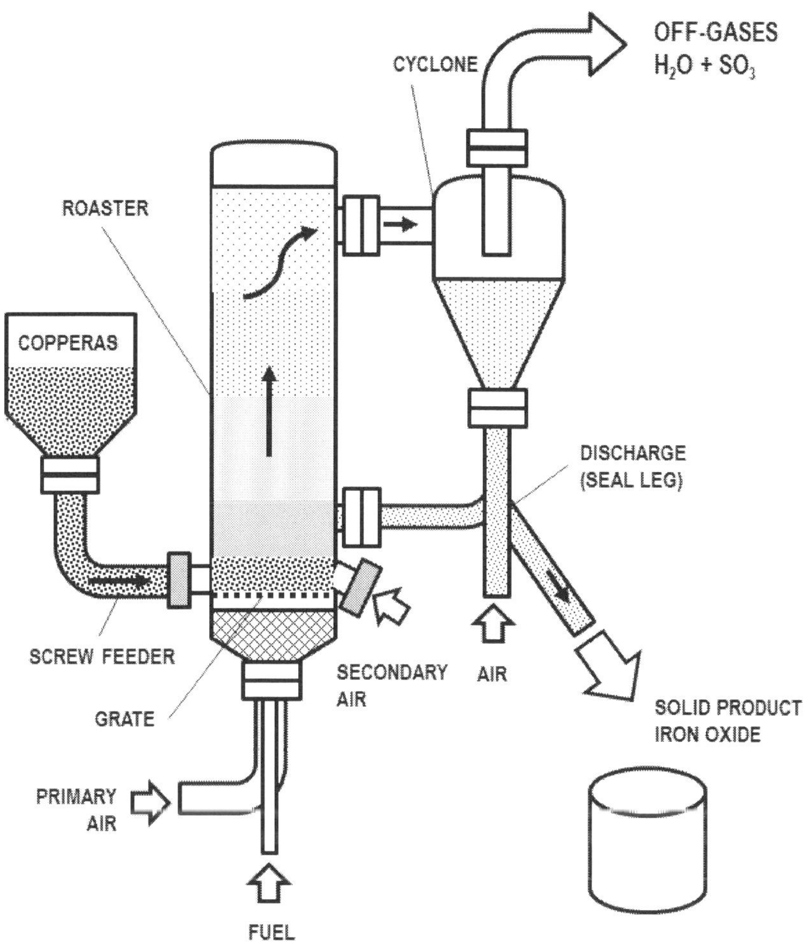

Figure 38 – Schematic of the roasting process for ferrous sulfate monohydrate.

The first step is the dehydration of copperas (FeSO$_4$·7H$_2$O) in a rotary kiln to yield ferrous sulfate monohydrate.

$$FeSO_4 \cdot 7H_2O(s) + heat \rightarrow FeSO_4 \cdot H_2O(s) + 6H_2O(g)$$

[162] KLADNIG, W.F. (2010) Acid recycling in steel pickling plants: state-of-the-art and new developments in environmental protection. *International Journal of Environment and Waste Management*, **5**(3-4)368-378.

Granules of the produced monohydrate are classified. The fine part of the monohydrate is sold while the coarse part of the monohydrate is roasted.

The roasting process uses the heat provided by the combustion of a fuel to decompose the iron sulfates. In order to ensure complete decomposition of the iron sulfate and combustion of organic impurities, the decomposition process is carried out at a temperature of more than 1000°C under oxidizing conditions. In practice, the heat required for this strongly endothermic process is supplied by means of hot flue gases from the combustion of fossil fuels or sulfur-containing fuel. A large portion of the additional heat can be recovered when the gases leaving the reactor are cooled from 1000°C to about 350°C. Depending on the intended use of the recovered energy, different waste-heat systems may be employed.

The thermal decomposition yields iron oxides and sulfur dioxide gas. Afterwards, the sulfur dioxide (SO_2) needs to be converted further to sulfur trioxide (SO_3) by means of a mandatory nearby and expensive sulfuric acid plant.

$$4FeSO_4 \cdot H_2O(s) + O_2(s) + heat \rightarrow 2Fe_2O_3(s) + 4SO_3(g) + 4H_2O(g)$$

The calcined iron oxide product is suspended in water. The residual soluble salts, mostly ferric and ferrous sulfates are washed out. All the waste water is treated in the neutralization plant. The classified slurry of iron oxides is dewatered and dried in a rotary dryer. The dried iron oxide pigment is ground, homogenized and packed into 25-kg paper bags or 1,000-kg super sacks. At present, the European plants using this technology are listed in Table 72 [163].

[163] HAMMERSCHMIDT, J.; and WROBEL, M. (2009) Decomposition of metal sulfates: a SO_2-source for sulfuric acid production. *Sulphur and Sulphuric acid Conference 2009*, The Southern African Institute of Mining and Metallurgy (SAIMM), Johannesburg, South Africa, May 4th-6th, 2009.

Table 72 – European titanium plants roasting copperas and pickling liquors

Company	Location	H_2SO_4 capacity (tonnes/day)	Feedstock(s)
Sachtleben Chemie GmbH	Sachtleben, Germany	2,000	Filter salt, spent acid
Rohm & Huls	Worms, Germany	650	Spent pickling liquors
Tronox	Krefeld, Germany	600	Filter salt, pyrite, spent acids
Rohm & Huls	Wesseling, Germany	400	Spent pickling liquors
Kronos (National Lead)	Leverkusen, Germany	435	Spent acid
Huntsman Tioxide	Calais, France	270 (Closed)	Filter salt, coal
Cinkarna	Celje, Slovenia	235 (Dismantled)	Filter salt, pyrite

The total amount of copperas roasted worldwide annually is estimated to 450,000 tonnes. The iron oxide product is used to supply the manufacture of iron oxide pigments, and about 120,000 tonnes to supply the manufacture of magnetic oxides.

Unfortunately, because nowadays the market for magnetic storage devices such as recording tapes or floppy disks is in severe decline and the market of iron oxide pigment is very limited, the commercial value of the iron oxides drastically decrease making the process less profitable and some plant must blend the copperas with spent pickling liquors to keep the installation running.

The specific energy consumptions, the amount of various emissions and pollutants along with other environmental impact factors of the roasting processes are listed in Table 73 [164]. The figures show the strong energy consumption of the thermal roasting process.

[164] COLLECTIVE (2006) *Technical Brochure on Copperas and Related Products, Chapter 7.* Published by the European Chemical Industry Council (CEFIC)-Titanium Dioxide Manufacturers Association (TDMA), Brussels, Belgium.

Specific energy consumption per unit mass of copperas	2,234 kWh/tonne
Specific energy consumption per unit mass of solid product	7,777 kWh/tonne
Specific energy consumption per unit mass of sulfuric acid produced (93%) (**)	5,895 kWh/tonne
Carbon dioxide emission per unit mass of copperas (***)	741 kg CO_2/tonne
Wastewater released per unit mass of copperas treated (*)	7,759 kg/tonne
Total pollutants emission (e.g., SO_x, NO_x, dust and particulates) to air per unit mass of copperas	12 kg/tonne
Notes: (*) wastewaters by-produced during the washing of the calcined iron oxides; (**) includes the specific energy required by the vacuum thermal evaporation to concentrate the acid from 30% until 93%. (***) Based on actual fuel consumption assuming a combustion thermal efficiency of 60%.	

11.1.6 Converting Copperas to Epsom Salt

Magnesium sulfate heptahydrate, $MgSO_4.7H_2O$, also called ***Epsom salt*** in the trade was produced from spent copperas by reacting low iron grade dolime (i.e., burned dolomite: $MgO + CaO$) with a concentrated aqueous solution of ferrous sulfate under aerated conditions. The chemical reaction is as follow:

$$2FeSO_4 + CaO + MgO + 2H_2O + O_2 = MgSO_4 + CaSO_4.2H_2O + 2FeO(OH)(s)$$

The hot water leaching of the sludge produced yield a saturated brine of magnesium sulfate while gypsum and the freshly precipitated ferroso-ferric hydroxide are separated by gravity settling and cross flow filtration.

The clear pregnant leach solution containing between 350-400 g/L of $MgSO_4$ is concentrated by multi-stage vacuum evaporation, and the crystallized Epsom salt is recovered by centrifugation, followed by mild drying to obtain dry crystals.

The process was performed commercially at the beginning of the 1900s in the USA by the *Ohio Chemical & Manufacturing Company* [165].

[165] CLARKE, G. (1913) *Process of making Epsom salts.* U.S. Patent 1,112,770 A (Ohio Chemicals & Manufacturing Company), October 6th, 1914.

11.1.7   Electrowinning Iron (FerWIN Process)

The patented ***FerWIN™ process*** [166] produces pure electrolytic iron (99.995 wt.% Fe) or iron-rich alloys (e.g., Fe with Ni, Cr, Mn, and V) from industrial wastes, and liquid effluents containing iron sulfate(s) such as copperas, spent pickling liquors (SPLs), pregnant leach solutions (PLS) while regenerating concentrated sulfuric acid, and releasing pure oxygen gas. A schematic flow diagram of the entire process is depicted in Figure 39.

Pilot testing conducted on various effluents originating from the chemical and metallurgical industries, confirmed the excellent cathode current (faradaic) efficiency of 98%, the high space time yield (i.e., throughput) up to 25 kg per day and per m^2 of active cathode surface, and the low specific energy consumption of only 2.9 MWh per tonne of metallic iron compared to 6.38 MWh per tonne for steel when using the conventional blast furnace. The latter figure makes the FerWIN™ technology a true green and zero-carbon ironmaking process.

From an environmental standpoint, the FerWIN™ process exhibits also a carbon negative footprint as it releases pure oxygen gas to the atmosphere compared to conventional steelmaking that releases on average 1.8 tonne CO_2 per tonne of steel. Thus in addition to monetize the iron values and to regenerate the sulfuric acid, it can be used to offset carbon taxes.

The patented technology which is now granted and enforced globally (i.e., Canada, India, Brazil, China, Japan, South Africa, and the European Union) can be suitable to be used in the following three industrial scenarios:

[1] When large amounts of metallurgical wastes, and chemical effluents containing iron sulfate(s) are by-produced annually at a given production facilities posing environment issues, it allow avoiding the disposal and landfill costs;

[2] When the supply of sulfuric acid is rather difficult and because the sulfuric acid consumed by a process ends-up as iron sulfate(s) in the effluents, the electrowinning process allow regenerating efficiently all the sulfuric acid values that are recycled back in the upstream process thus reducing the working capital and operating costs significantly;

[3] Finally, the process allow to monetize all the iron values contained in the wastes and to obtain the maximum value for the iron product as the commercial price for pure metallic iron is by far higher than the price for any of the other iron products generated using other routes (e.g., copperas, hematite).

---

[166] CARDARELLI, F. (2016) *Electrochemical process for the recovery of metallic iron and sulfuric acid values from iron-rich sulfate wastes, mining residues, and pickling liquors.* Canadian Patent CA 2,717,887 C (Electrochem Technologies & Materials Inc.), June 14th, 2016.

Despite the technology is not yet commercial, it has been tested at the pilot scale [1:50] and it is now technically proven, de-risked, and the favorable costs and benefits analysis suggest that combining the sulfation of bauxite residues and the electrowinning of iron could represent a possible route for neutralizing, dewatering, recycling and valorizing red mud and bauxite residues. For instance, the operating cost of producing pure electrolytic iron can be as low as $315.00 per tonne in certain locations. This is particularly true in locations having a short supply of sulfuric acid while having direct access to affordable nuclear or hydroelectricity such as Quebec, and Scandinavia.

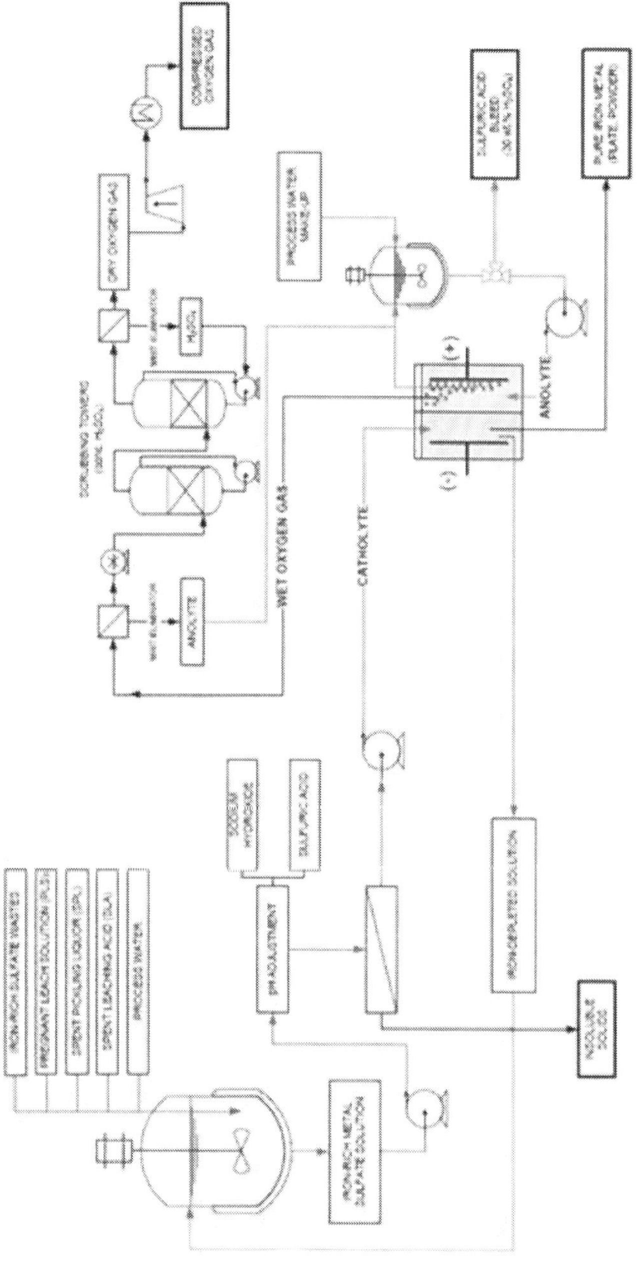

Figure 39 – Schematic of the FerWIN process for electrowinning iron

11.2 Gypsum

Because of the ubiquitous presence of calcium oxide in several raw materials that are sulfated such as phosphate rock, bauxite, and slags, most of the calcium reacts with the sulfuric acid and ends up as either gypsum ($CaSO_4.2H_2O$) or anhydrite ($CaSO_4$) depending on the amount of water present, the operating temperature, and concentration of acid.

Moreover, the neutralization of spent acid and acidic liquids effluents by-produced during the sulfation titania or acid mine drainage (AMD) is usually conducted using quicklime (CaO), hydrated or slacked lime [$Ca(OH)_2$] or ground limestone ($CaCO_3$) also produce large amount of gypsum. Depending on the

11.2.1 Phosphogypsum

The so-called ***phosphogypsum*** is the calcium sulfate dihydrate formed as a by-product of the production of fertilizer from the acidulation phosphate rock (See Section 9.2). It is mainly composed of gypsum largely contaminated by free silica, fluorides, and other deleterious contaminants such as iron, aluminum, cadmium, barium, lead, and radium.

On average, about five tonnes of phosphogypsum are generated per tonne of phosphate and phosphoric acid production Globally, estimated generation of phosphogypsum is huge ranging from 96 to 275 million tonnes with 160 million tonnes [167].

The largest tailings are all located nearby the phosphate fertilizer plants in China, the USA, North Africa, Russia, Brazil, and Middle East. Moreover, it is reported that Florida tailings represent up to forty percent of the worldwide phosphogypsum by-produced.

Although clean gypsum is a widely used material in the construction industry, phosphogypsum is usually not suitable as the content of silica and iron precludes its use commercially, especially the latter imparting a tint affecting the brightness and whiteness compared to genuine gypsum. Another major reason is its weak radioactivity caused by the presence of naturally occurring radionuclides such as uranium (238-U) and thorium (232-Th), and their daughter isotopes radium (226-Ra), radon (222-Rn), and polonium (Po).

Thus phosphogypsum is considered from a regulatory standpoint according to the *International Atomic Energy Agency* (IAEA) as a ***technologically enhanced naturally occurring radioactive material*** (TENORM) and it is thus

---

[167] COLLECTIVE (2013) *Radiation Protection and Management of NORM Residues in the Phosphate Industry.* Safety Report Series 78, International Atomic Energy Agency (IAEA), Vienna, Austria.

---

prevented to be used as building materials [168]. Therefore, it is simply landfilled forming large tailing hills in the landscape (e.g., Florida, Tunisia).

11.2.2 Titanogypsum

Similarly, during the production of titanium dioxide by the sulfate route (see Section 9.1), very large amount of the so-called ***titanium gypsum*** or in short ***titanogypsum*** is produced following to the neutralization of excess weak acid with quicklime or slacked lime [169]. As for phosphogypsum, the final product is always contaminated with silica, fluorides, organic matters, and other alkalis (e.g., Na, Mg) rendering the gypsum improper for being used commercially.

In mainland China where the amount of titanium gypsum landfilled in tailings is rather significant due to the plethora of pigment plants producing titanium dioxide by the sulfate route some researchers from *Tsinghua University Guangdong* and *Tongke Environmental Protection Technology Co., Ltd.* considered to use titanium gypsum as raw material to prepare a granular soil material exhibiting a porous and aerated structure ideal for water retention and suitable to be used in arid regions [170].

During the preparation process the titanium gypsum is calcined at high-temperature to drive-off bound water and thus convert the gypsum into anhydrite with a greater reactivity towards the other ingredients. Afterwards, the silica and alumina content from a plasticizer and a pore-forming agent are subjected to a gelatinization reaction so as to enhance the strength of the granular soil. Other use such as in cement as stabilized crushed stone was also investigated by Zhao et al. [171].

11.2.3 Recovery of Sulfuric Acid from Gypsum

Sulfuric acid can be produced from gypsum by the ***Müller-Kühne process***. The sulfuric acid can be obtained from the sulfur dioxide evolved during the high temperature roasting of gypsum with silica sand at 1400°C under forced aerated conditions according to the following chemical reaction:

[168] MAZZILLIA, B.P.; CAMPOSA, M.P.; NISTIA, M.B.; SAUEIAA, C.H.R.; and MÁDUARA, M. F. (2020) Radiological implications of using phosphogypsum as building material: a case study of Brazil. *Brazilian Journal of Radiation Sciences*, **8**(1)1-29.

[169] ZHANG, Y.; WANG, F.; HUANG, H.; GUO, Y.; LI, B.; LIU, Y.; and CHU, P.K. (2016) Gypsum blocks produced from TiO₂ production by-products. *Environmental Technology* 37(**9**)1094–100.

[170] KAI, Z.; JIESHAN, C.; and MINGYUE, D. (2018) *Titanium gypsum granular soil and preparation method thereof.* Chinese Patent Application 2018/11122132.6, September 26th, 2018.

[171] ZHAO, Z.; LIU, H.; ZHAO, Y.; WANG, R.; LIU, G.; and YANG. Z. (2020) *Research on Road Performance of Titanium Gypsum Used in Cement Stabilized Crushed Stone.* 20th COTA International Conference of Transportation Professionals, CICTP, 2020: Advanced Transportation Technologies and Development-Enhancing Connections, American Society of Civil Engineers (ASCE), August 14-16, 2020, Xi'an, China.

$$CaSO_4(s) + SiO_2(s) = CaSiO_3(s) + SO_2(g) + 0.5O_2(g)$$

However, in practice, the above reaction is performed inside a rotary kiln and it is promoted by adding about 5-10% of pulverized coke to the gypsum charge in order to decrease the reduction temperature to 900°C while using clays as source of silica instead of quartz sand, an alumino-silicate **clinker** is by-produced with excellent cementiferous properties suitable for the cement industry [172].

Therefore, the above chemical reaction needs to be rewritten as it now occurs in three consecutive steps depending on the temperature region existing inside the kiln:

$$CaSO_4(s) + 2C(s) = CaS(s) + 2CO_2(g)$$

$$CaS(s) + 3CaSO_4(s) = 4CaO(s) + 4SO_2(g)$$

$$4CaO(s) + (Fe_2O_3 + Al_2O_3 + SiO_2) = Clinker$$

The industrial process was once used since 1918 by the German company *Bayer AG* in Leverkusen until 1990. In the process devised and implemented by the German engineering and construction firm *Lurgi GmbH*, the ground gypsum, usually either phosphogypsum or titanogypsum, was mixed with 5-10% pulverized coal, and clay material and feed to a 80-meter long brick-lined rotary kiln with about 3-meter inside diameter.

After passing through a dust collector to remove the dust and fine particulates, the off-gases consisting mainly of a mixture of air and sulfur dioxide (7 vol.% SO_2) were mixed with the off gases originating from the nearby copperas roasting plant before to be sent finally to the sulfuric acid plant using the direct contact process.

On the other hand, the molten clinker obtained was mixed with blast furnace slag to yield a Portland cement of excellent commercial value.

In addition to the five plants that have been built and operated in Germany and Austria, several other plants were built worldwide including the United Kingdom, France, the USA, Poland, South Africa, and China [173] but to our knowledge except in China none seems to have survived until today.

---

[172] LUDTKE, P.; and SILBERBERG, A.N. (1988) Is gypsum decomposition commercially ready now? AIChE Clearwater Convention, Clearwater, FL.

[173] SWIFT, W.M.; PANEK, A.F.; SMITH, G.W.; and VOGEL, G.J.; and JONKE, A.A. (1976) *Decomposition of Calcium Sulfate. A Review Literature.* Argonne National laboratory (ANL), Report ANL-76-122, Argonne, IL.

11.3 Sodium Sulfate (Glaubert's salt)

When sodium is present in significant concentration such as in bauxite residues, the sulfation produce large amount of **sodium sulfate dodecahydrate**, with chemical formula $Na_2SO_4.10H_2O$, CAS No. [7727-73-3], also called **Glaubert's salt** in the trade that crystallizes easily in the lines as the solubilities is not high and decreases with increasing temperature.

Several industrial activities produce waste sodium sulfates as the result of sulfation of Na-rich raw materials (e.g., bauxite residues, slags) or following the neutralization with soda ash of liquid effluents by-produced by sulfation. Because these wastes are controlled by stringent environmental regulations for wastewater concentrations of sulfate, the contaminated sodium sulfate needs to be disposed. However, several approach to recycle sodium sulfate are available one option is to reuse the sodium sulfate in other processes (e.g., steel pickling industry).

The world production of sodium sulfate decahydrate ranges between 5.5 to 6 million tonnes annually.

About one third of the world's sodium sulfate is produced as by-product of other processes in chemical industry. Most of this production is chemically inherent to the primary process, and only marginally economical. By effort of the industry, therefore, sodium sulfate production as by-product is declining.

In the fertilizer industry, sodium sulfate either in the form of mirabilite (Glaubert's salt) or a concentrated brine of pure sodium sulfate can be reacted with a brine containing potassium chloride or muriate of potash to form a double salt called **glaserite** with chemical formula $K_6Na(SO_4)_4$ that can be decomposed to yield potassium sulfate to produce potassium sulfate and sodium chloride. Worldwide there are only a few producers utilising this process such as K+S Kali in Germany, and *Rusal* in Russia. This method of production is the second greatest source of global supply at 25% to 30% with an average cost of production of $300 per tonne when compared with the Mannheim process described in Section 9.11.

12 Concentration of Sulfuric Acid

12.1 Evaporation under Vacuum

The concentration of weak and spent sulfuric acids is essentially performed industrially by evaporating the water content until the final concentration of sulfuric acid is obtained. Usually the feed input for the recovery of highly concentrated sulfuric acid must exhibit a mass percentage ranging from 70 to 80 wt.% H_2SO_4 but lower concentration are more and more common especially when dealing with weak and spent acids from the metallurgical industries. However, during the evaporation of the latter, the high level of inorganic impurities yield gradually a suspension of concentrated sulfuric acid with anhydrous metal salts that crystalize out and that must be removed prior the final concentration stage usually by gravity settling or centrifugation.

The typical equipment to perform this type of concentration is called an **evaporator** and sometimes **concentrator**. Usually, either steam, or natural gas heaters, or even quartz or silicon carbide immersed electrical heaters are used to supply the necessary sensible and latent heats.

For practical reasons, the overall evaporation process is always conducted under negative gauge (reduced) pressure (i.e., vacuum) in order to decrease the boiling temperature. Actually, because the thermal efficiency of the evaporator is directly related to the difference between the temperature of the condensing steam and the boiling temperature of the feed (i.e., diluted sulfuric acid), decreasing the absolute pressure inside the vapor space well below atmospheric pressure will reduce the boiling point of the acid significantly and thus enable the use of steam as source of heat. Secondly, this is mandatory to minimize the corrosion of the construction materials under such harsh operating conditions encountered especially in the last stages where the hot and strong concentrated acid poses serious challenges even to well-known corrosion materials such as tantalum, quartz, and silicon carbide (see Section 6).

The optimum reduced pressure is chosen based on the temperature difference between the superheated steam and the boiling point of sulfuric acid. Most importantly as reported in Section 5.5 when evaporating sulfuric acid with a mass percentage below 78 wt% H_2SO_4 only water predominates in the vapor phase while above that threshold some sulfuric acid and sulfur trioxide (SO_3) are also present in the vapor phase.

Moreover, as already discussed in Section 3.2, the sulfuric acid-water system exhibits under atmospheric pressure (101.325 kPa) a negative (maximum) azeotrope with a concentration of 98.5 wt.% H_2SO_4 with a boiling temperature of 338°C. Therefore, it is not possible to obtain sulfuric acid with a final concentration of 100 wt.% H_2SO_4 only by thermal evaporation.

Actually, in practice, a first evaporation from 78 wt% H_2SO_4 up to a maximum sulfuric acid concentration of 88-90 wt.% H_2SO_4 is conducted by forced circulation inside a triple-effect evaporator due to the low partial pressure of sulfuric acid in the vapor phase.

Afterwards, a second evaporation is performed under a lower pressure inside a horizontal evaporator equipped with tantalum immersed heaters. At that point, a mist eliminator is installed in order to reduce the proportion of sulfuric acid mist entrained in the vapor phase. This last stage yields the strong concentrated sulfuric acid with 96 wt.% H_2SO_4. Finally, if a higher strength is needed either oleum or sulfur trioxide (SO_3) are injected to adjust the final concentration.

In the following sections, we will detail the basic equations involved during the concentration of sulfuric acid with a single-effect evaporator with a practical example to support the calculations.

12.1.1 Single-Effect Evaporation

In this section, a simplified mathematical approach will be used for concentrating directly diluted sulfuric acid (25 wt.% H_2SO_4) up to the concentrated sulfuric acid (93 wt.% H_2SO_4). Despite, not being realistic, these calculations will provide the baseline of a worst case scenario for the performances of a single-effect evaporator.

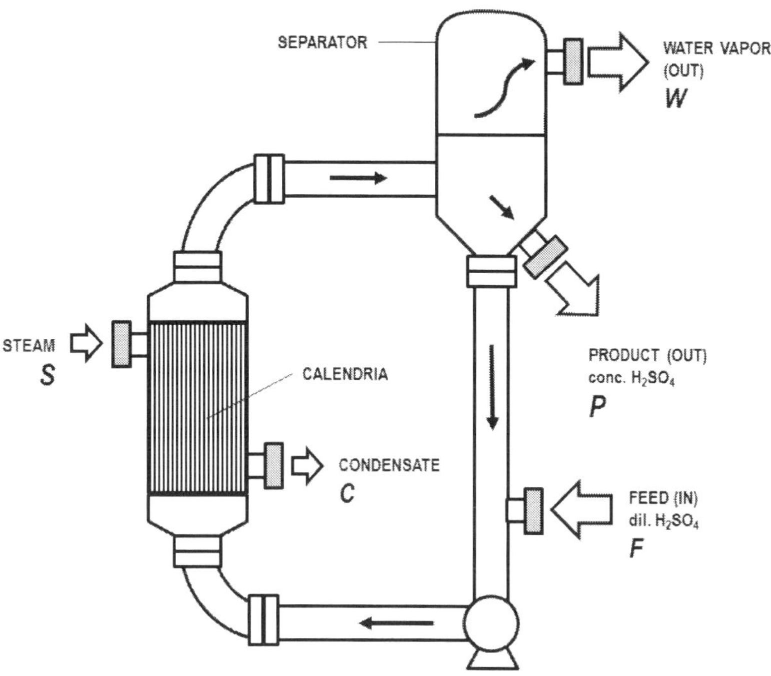

Figure 40 – Single-effect forced circulation evaporator

In the following calculations, the acid concentration plant annual nameplate capacity is taken as 25,000 tonnes during a 350 operating-days period. Moreover, an absolute pressure of 10 kPa (1.450 psia) is maintained inside the vapor space.

Table 74 – Sulfuric acid concentration plant capacity

PLANT NAMEPLATE CAPACITY			
Annual production capacity (93.3 wt.% H₂SO₄	P	25,000 tonnes/year	27,558 tons/year
		71,429 kg/day	157,473 lb/day
		2,976 kg/hour	6,561 lb/hour
Process losses	λ	10%	
Production period		350 working days	8,400 working hours

12.1.2   Properties of the Dilute Sulfuric Acid (Feed)

The feed that consists of a diluted sulfuric acid solution with a mass percentage w_F of 25 wt.% H₂SO₄ enters the evaporator at 50°C (122°F). Its mass density is 1,178.3 kg/m³ (73.56 lb/ft³) when measured at 20°C (68°F). The mass flow rate of the feed, F, is 266.57 tonnes per day (587,689 lb/day).

At the absolute pressure existing in the vapor space, according to the *International Steam Tables* [174, 175], pure water boils at 45.8°C (114.48°F). Under this same reduced pressure and from the data obtained from the Dühring's rules applied to the water-sulfuric acid solutions (see Section 5.7), the extrapolated boiling point of the dilute acid solution is 50.1°C (122.2°F).

Table 75 – Vacuum operations conditions

VACUUM EVAPORATION CONDITIONS (SINGLE EFFECT)			
Absolute residual pressure (vapor space)	P_{VS}	10.0 kPa	1.450 psia
Boiling point of pure water @ P_{VS}	bp (water)	45.8°C	114.5°F

Hence the ***boiling point elevation*** (*BPE*) or ***boiling point rise*** (*BPR*) of the dilute solution is only +4.2°C (7.5°F). The specific enthalpy of the feed obtained from the enthalpy-concentration diagram (Section 8.7) is +1.11 kJ/kg (+0.48 Btu/lb).

Table 76 – Dilute sulfuric acid (Feed) properties

FEED (Diluted sulfuric acid)			
Mass density @ 20°C	ρ_F	1.178.3 kg/m^3	73.56 lb/ft^3
Specific heat capacity of feed	c_F	3,485 J/kg/K	0.832 Btu/lb/°F
Concentration of sulfuric acid	w_F	294.58 g/L	25.0%
Temperature (Feed)	T_F	50.0°C	122.0°F
Temperature (Pre-Heated)	T_{PH}	50.0°C	122.0°F
Boiling point @ P_{VS}	bp	50.0°C	122.0°F
Boiling point elevation	BPE	4.18°C	7.5°F
Specific enthalpy content of feed	h_F	1,116 J/kg	0.48 Btu/lb
Specific enthalpy of pre-heated feed	h_{PH}	1,116 J/kg	0.48 Btu/lb
Heat rate for pre-heating the feed	Q_{PH}	0.0 kJ/day	0.0 MMBtu/day
Feed mass flow rate	F	93,300 tonnes/year	102,846 tons/year
		266,571 kg/day	587,689 lb/day
		11,107 kg/hour	24,487 lb/hour
Volume flow rate	V_F	226.234 m^3/day	59,765 gal/day

12.1.3 Properties of the Concentrated Sulfuric Acid (Product)

The concentrated sulfuric acid or product leaves the evaporator as a concentrated sulfuric acid solution with a mass percentage, w_P, of 93.3 wt.% H_2SO_4 and a mass density of 1,828 kg/m^3 (114.1 lb/ft^3) when measured at 20°C (68°F). Therefore, the mass flow rate of the product, P, is simply given by the following equation: $P = F (w_F/w_P) = 71.43$ tonnes per day (157,473 lb/day).

[174] Collective (1967) *Steam Tables. Properties of Saturated and Superheated Steam – From 0.08865 to 15,500 lb per sq. in. absolute pressure*. Combustion Engineering, Inc., Winsor, CT.

[175] WAGNER, W.; and KRETZSCHMAR, H.-J. (2008) - *International Steam Tables. Properties of Water and Steam based on the Industrial Formulation IAPWS-IF97, Second Edition*. Springer-Verlag, Heidelberg, 392 pages.

For such elevate concentration, the specific enthalpy of dilution is non negligible, and thus the concentrated acid boils at a temperature much higher than that of pure water. Actually, under the absolute pressure existing in the vapor space, according to Dühring's rules, the concentrated sulfuric acid boils at 196.6°C (385.9°F). Hence the boiling point elevation of the concentrated acid is 150.8°C (271.4°F). The specific enthalpy of the feed obtained from the enthalpy-concentration diagram (Section 8.7) is +170.5 kJ/kg (+73.3 Btu/lb).

Table 77 – Concentrated sulfuric acid (Product) properties

PRODUCT (Concentrated sulfuric acid)			
Mass density @ 20°C	ρ_P	1,828 kg/m³	114.1 lb/ft³
Specific heat capacity of product	c_P	1,585 J/kg/K	0.379 Btu/lb/°F
Concentration of sulfuric acid	w_P	1,706 g/L	93.3%
Boiling point @ P_{VS}	bp	196.6°C	385.9°F
Boiling point elevation	BPE	150.8°C	271.5°F
Specific enthalpy content of product	h_P	170,449 J/kg	73.3 Btu/lb
Product mass flow rate	P	25,000 tonnes/year	27,558 tons/year
		71,429 kg/day	157,473 lb/day
		2,976 kg/hour	6,561 lb/hour
Daily volume flow rate	V_P	39.065 m³/day	10,320 gal/day

12.1.4   Evaporated Water Vapor (Vapor)

The mass flow rate of water vapor (V) leaving the head space as superheated steam is given by the difference between the mass flow rates of feed and product: $V = F - P = 195.14$ tonnes per day (430,216 lb/day).

The vapor leaves the vapor space at 196.6°C (385.9°F) with a superheat of 150.8°C (271.5°F) and with a latent enthalpy of vaporization of 2.402 MJ/kg (1,033 Btu/lb).

Table 78 – Water vapor (Vapor) properties

VAPOR (SUPERHEATED STEAM)			
Temperature of steam leaving the evaporator	T_V	196.6°C	385.9°F
Saturation temperature @ P_{VS}	T_S	45.8°C	114.5°F
Superheat of vapor	Sh	150.8°C	271.5°F
Specific heat capacity of vapor	c_V	1,888 J/kg/K	0.451 Btu/lb/°R
Specific latent enthalpy of vaporization	L_V	2,402,758 J/kg	1,033.0 Btu/lb
Specific enthalpy of vapor	h_V	2,869,587 J/kg	1,233.7 Btu/lb
Specific volume of vapor	v_g	0.028 m³/kg	2.17 ft³/lb
Vapor mass flow rate	V	68,300 tonnes/year	75,288 tons/year
		195,143 kg/day	430,216 lb/day
		8.131 kg/hour	17,926 lb/hour

12.1.5   Heat and Mass Balances

Based on the mass and energy balance existing inside the evaporator, the total enthalpy contained in the feed plus the heat of evaporation provided must be equal to the total enthalpies contained in the product and the vapors, as follows:

Overall mass balance: $\qquad F = P + V$

Mass balance of sulfuric acid: $\qquad w_F\, F\cdot = w_P\, P$

Mass balance of water: $\qquad (1 - w_F)F = (1 - w_P)P + V$

Overall energy balance: $\qquad h_F\, F\cdot + Q_E = h_P\, P\cdot + h_V\, V$

Where Q_E is the heat flow rate expressed in W (Btu/h) required to concentrate the diluted acid, F, P, and V, and h_F, h_P and h_V, the mass flow rates and specific enthalpies of the feed, product, and vapor respectively.

12.1.6 Overall Heat Demand

Hence, the overall heat demand to be supplied is given by the algebraic difference between the enthalpy content of the superheated water vapor leaving the evaporator plus the enthalpy of the product minus the enthalpy of the feed:

$$Q_E = h_V\,(F - P) + h_P\, P - h_F\, F$$

In the particular case of sulfuric acid solutions, because of the considerable release of heat upon mixing with water, it is not possible to neglect the specific enthalpy of dilution and, the specific enthalpy of the feed and product must be determined from the enthalpy-concentration diagram for water-sulfuric acid mixtures (see Figure 14).

$$h_F = [h_F(T_F) - h_F(T_R)]$$

$$h_P = [h_P(T_P) - h_P(T_R)]$$

The reference temperature, T_R, can be chosen as any convenient value such as, for instance, as the temperature of the feed, T_F.

Regarding the vapor, it is important to consider the latent enthalpy of vaporization, L_v, also denoted h_{fg} in the steam table the sensible heat of the vapor (c_v) itself if superheated, and the sensible enthalpy of the condensate.

$$h_V = [h_V(T_V) - h_V(T_R)] \qquad = c_V\cdot(T_V - T_S) + L_v + c_C\cdot(T_S - T_R)$$

Where c_V, and c_C are the isobaric specific heat capacities, in J/kg/K (Btu/lb/°F), of superheated water vapor, and the water condensate respectively.

In the absence of measured data, the latent enthalpy of vaporization in kJ/kg and the latent entropy of vaporization, S_V, s_{fg}, in kJ.kg^{-1}.K^{-1}, can be calculated at an absolute temperature T in K, with a good accuracy using the empirical equation:

$$L_v \text{ (kJ/kg)} - h_{fg} \text{ (kJ/kg)} = 3{,}328.4 - 2.9\,T(\text{K})$$

$$S_v \text{ (kJ.kg}^{-1}\text{K}^{-1}) = s_{fg} \text{ (kJ.kg}^{-1}.\text{K}^{-1}) = 3{,}328.4/T(\text{K}) - 2.9$$

Similarly the saturation pressure in kPa is obtained for an absolute temperature in K by:

$$P_{sat} = 101.325 \exp(56.54 - 7{,}212/T - 6.285\,\ln T)$$

Therefore, the heat to be supplied to the evaporator is given by the equations:

$$Q_E = (F - P)\cdot[h_V(T_V) - h_V(T_R)] - F\cdot[h_F(T_F) - h_F(T_R)] + P\cdot[h_P(T_P) - h_P(T_R)]$$

$$Q_E = (F - P)\cdot[\,c_V\cdot(T_V - T_S) + L_v + c_C\cdot(T_S - T_R)] - F\cdot[h_F(T_F) - h_F(T_R)] + P\cdot[h_P(T_P) - h_P(T_R)]$$

From the *International Steam Tables* and the enthalpy-concentration diagram, the specific enthalpies content of the feed, product, and vapor are:

Feed (25 wt.% H_2SO_4, 122°F) $\qquad\qquad h_F = -1.12$ kJ/kg (+0.48 Btu/lb)

Product (93.3 wt.% H_2SO_4, 386°F) $\qquad h_P = 170.5$ kJ/kg (+73.3 Btu/lb)

Vapor (1.450 psia, 386°F) $\qquad\qquad\quad h_V = 2{,}870$ kJ/kg (+1,234 Btu/lb)

Therefore, the overall and actual heat flow rate required is equal to:

$$Q_E = +6{,}619 \text{ kW} = +542 \text{ MMBtu/day}$$

Table 79 – Summary of heat flow rates and energy balance

HEAT FLOW RATE and POWER DENSITY (SINGLE EFFECT)			
Sensible heat rate to pre-heat the feed	Q_{PH}	0 kW	0 MMBtu/day
Sensible heat rate to rise the feed (from T_R)	Q_H	0 kW	0 MMBtu/day
Heat rate to generate the water vapor	Q_V	6,481 kW	531 MMBtu/day
Overall heat flow rate (neglecting heat dilution)	$Q_{PH}+Q_H+Q_V$	6,481 kW	531 MMBtu/day
Overall heat flow rate (with heat of dilution)	$Q_{PH} + Q_E$	6,619 kW	542 MMBtu/day
Specific energy for evaporation (Feed)		2,145 kJ/kg	922 Btu/lb
Power density (Theoretical)		2,528 kW/m^3	32,649 Btu/h/gal

12.1.7 Steam Economy

In order to perform the concentration of the sulfuric acid, one common industrial solution is to supply the heat required by mean of saturated steam to a tube and shell heat exchanger called the calandria.

The heat flow rate, q_S, in W (Btu/h), supplied by the condensing steam transferred to the feed through the walls surface of the heat exchanger is given by the product of the mass flow rate of the condensing steam, S, in kg/s (lb/h) minus the mass flow rate of the amount of steam lost by venting the non-condensable, denoted u, in kg/s (lb/h), and the difference between the specific enthalpy content of the steam and hot water condensate, C, in J/kg (Btu/lb):

$$Q_S = (S - u)\cdot(h_S - h_C) = S\cdot\Delta h_v + C\cdot c_p(T_S - T_P) + u\,(h_S - h_u) \qquad \text{with} \qquad u = a\,S$$

Therefore the total enthalpy content of the saturated steam consists mostly to its specific latent enthalpy of vaporization plus eventually the sensible heat released from the condensate. The term $u\,(h_S - h_u)$ corresponds to the heat loss during sub-cooling of the vent stream.

$$Q_S = (S - u)\,[L_S + c_C\cdot(T_S - T_R)]$$

If we use saturated steam with a condensing or saturation temperature (T_S) of 207°C (405°F), according to the *International Steam Tables*, its absolute pressure is 1,804 kPa (262 psia), and its specific latent enthalpy of vaporization, L_S, is 821 Btu/lb (1,911 kJ/kg) and the specific enthalpy is 1,202 Btu/lb (2,795 kJ/kg).

Therefore, the mass flow rate of saturated steam, S, in kg/day (lb/day), required taking into account the processing losses, λ, and the amount of steam vented, $u = aS$, is given by the following equation:

$$S = (1 + \lambda)(Q_E/h_S)/(1 - a)$$

Thus in this case, with a processing loss of 10%, the mass flow rate of superheated steam is 227 tonnes per day (501,114 lb/day).

Table 80 – Saturated steam demand

STEAM DEMAND (SATURATED STEAM)			
Steam condensing temperature	T_S	207°C	405°F
Steam pressure (absolute)	P_S	1,804 kPa	261.66 psia
Sensible heat released from the condensate	Q_S	884,811 J/kg	380.4 Btu/lb
Specific latent enthalpy of steam	L_S	1,910,577 J/kg	821.4 Btu/lb
Specific enthalpy content of steam	h_S	2,795,387 J/kg	1,201.8 Btu/lb
Mass fraction of steam vented	a	1.0%	
Daily mass flow rate of steam (no losses)	$S = Q/h_S(1-a)$	206,638 kg/day	455,559 lb/day
Daily mass flow rate of steam (losses)	$S^* = \lambda S$	227,302 kg/day	501,114 lb/day
Hourly mass flow rate of steam	S	9,471 kg/hour	20,880 lb/hour
Daily mass of steam vented for noncondensable	$u = aS$	2,066 kg/day	4,556 lb/day

12.1.8 Figures of Merit

As the purpose of a sulfuric acid concentration plants is to concentrate acids with a maximum thermal efficiency and a minimum capital investment (CAPEX) and operating cost (OPEX), it is important to introduce several important parameters that are used to assess the performances of the evaporator. These are:

The ***evaporator capacity*** (*EC*), the rate of vapor removal, is equal to 195 tonnes per day (17,926 lb/h).

The ***steam economy*** (*SE*), usually defined as the mass ratio between the water vapor evaporated per unit mass of steam consumed. In this case the vapor-to-steam mass ratio, is equal to 0.859 lb of vapor/lb of steam, that is, 85.6%.

The ***specific heat evaporation feed*** 2,145 kJ/kg (922 Btu/lb).

The ***steam consumption*** (*SC*), that is the mass flow rate of steam consumed daily, is equal to: 9,471 kg per hour (20,880 lb/hour).

The ***power density***, that is, the amount of heat required per unit volume of feed. In our case, the thermal power density is equal to 2,528 kW/m^3 (32,649 Btu/h/gal).

Table 81 – Figures of merit (single effect)

FIGURES OF MERIT (SINGLE EFFECT)			
Total energy required		2.00E+11 kJ	1.90E+11 MMBtu
Specific energy for evaporation (Feed)		2,145 kJ/kg	922 Btu/lb
Power density required		2,528 kW/m^3	32,649 Btu/h/gal
Evaporator capacity (rate water evaporated)	EC	8,131 kg/hour	17,926 lb/hour
Steam consumption (rate of steam consumed)	SC	9,471 kg/hour	20,880 lb/hour
Steam economy (vapor-to-steam mass ratio)	SE	0.859 kg /kg	0.859 lb/lb

12.1.9   Heat Transfer Calculations

The heat flow rate, *q*, expressed in W (Btu/h) across the walls of the heat exchanger is given by the following well-known equation used in heat transfer calculations:

$$q_{HX} = UAF\,\Delta T_{LMTD}$$

With U being the ***overall heat transfer coefficient***, in $W.m^{-2}K^{-1}$ ($Btu/h/ft^2/°F$), A the ***exchange surface area*** in m^2 (ft^2), and ΔT_{LMTD} the logarithmic mean temperature difference, in K (°F), and finally F is a dimensionless ***correction design factor*** related to the type and configuration of the heat exchanger.

In the particular case of the tube and shell heat exchanger discussed previously in the isothermal condensing of saturated steam with forced counter flow circulation, one shell pass for the condensing steam, and an even number of tube passes for the feed, the logarithmic mean temperature difference is as follows:

$$\Delta T_{\mathrm{LMTD}} = (T_P - T_F)/\ln[(T_S - T_F)/(T_S - T_P)] = 54 \text{ K } (97^\circ\mathrm{R})$$

Therefore, assuming an overall heat transfer coefficient of 2,500 W.m^{-2}.K^{-1} (440 Btu/h/ft^2/°F) and an F-factor equal to unity based on the following R-factor and P-factors definitions:

$$R = (T_S - T_S)/(T_P - T_F) = 0 \text{ and } P = (T_P - T_F)/(T_S - T_F) = 0.93$$

Therefore, the total surface area required for the proper heat transfer is equal to 7,359 m^2 (79,216 ft^2).

Table 82 – Sizing of the tube and shell heat exchanger

SIZING - Tube and Shell Heat Exchanger (Counterflow, one shell pass, even tube passes)			
Overall heat transfer coefficient (Forced convecti	U	2,500 Wm^{-2}K^{-1}	440 Btu/h/ft^2/F
Logarithmic temperature difference	$LMTD$	54°C	97°F
Correction design factor	F-factor	1.00	
Temperature of product	T_P	197°C	386°F
Steam condensing temperature	T_S	207°C	405°F
Feed inlet temperature	T_F	50°C	122°F
R-factor	R	0.00	
P-factor	P	0.93	
Exchange surface area	A	7,359 m^2	79,216 ft^2

12.1.10 Natural Gas Heater

If we need to use a natural gas heater for pre-heating the feed and the steam boiler, the overall thermal energy required must take into account the combustion efficiency of the gas burner, denoted e_B, and the heat transfer efficiency e_T. Therefore, the thermal energy content of the natural gas is calculated from the theoretical heat flow rate corrected by the combined efficiencies.

$$Q_{\mathrm{thermal}} = Q_E/(e_B \cdot e_T)$$

Assuming a thermal combustion efficiency of 83.4% and a heat transfer efficiency of 90%, the combined efficiency is 75.5% and hence the thermal energy required is 8.77 MW (718 MMBtu/day).

Table 83 – Pre-heater and natural gas boiler

SIZING - Pre-Heater and Natural Gas Boiler for Steam Generation			
Thermal combustion efficiency	e_B	83.9%	
Heat transfer efficiency	e_H	90.0%	
Combined efficiency	e_{BH}	**75.5%**	
Thermal energy (Total)	$Q_{thermal}$	8,765 kW	718 MMBtu/day
High heating value of fuel	HHV	37.26 GJ/m³	1.0 MMBtu/1000ft³
Consumption of natural gas (Total)	NG_T	20,326 m³/day	717,806 ft³/day

12.1.11 Triple-Effect Evaporator

12.1.11.1   Basics

Essentially, multiple-effect evaporators are designed to improve the energy economy of the evaporation process.

The basic principle is to reuse the heat released by the vapor condensation in one effect to provide heat for another effect. The vapor from the separator of the first effect enters the heat exchanger of the second effect, while live steam heats in the heat exchanger of the first effect. The steam and cooling water rates for the double effect unit are approximately 50% of those required for a single effect unit. For instance for an n-effect evaporator, the steam requirement is approximately $1/n$, but requires more heat exchange surface area and the additional heat-exchange surface required is approximately n times for the same output.

Table 84 – Simple, double and triple effect steam economy

Number of effects	Steam economy (lb steam/lb vapor)	Total running cost (as percentage of single effect evaporator)
Single-effect	1.16	100%
Double-effect	0.61	52%
Triple-effect	0.42	36%

12.1.11.2   Concentration of Sulfuric Acid from 25% to 70%

Triple-effects evaporators with forced circulation that recover heat by recycling steam have a proven track record for the concentration of sulfuric acid from 20 wt.% H_2SO_4 to 30 wt% H_2SO_4 up to a final concentration of 60 to 70 wt.% H_2SO_4. Due to great water evaporation, the condensation heat of the exhaust steam is almost fully recycled to heating in order to save energy. In this case, a forced circulation evaporator is the preferred choice to perform this task.

The overall heat transfer is ensured either by horizontal tube and shell heat exchangers (TS-HX) or plates and frame type heat exchangers both made of highly corrosion resistant tantalum metal (see Section 6.1.7.2) with a counter flow circulation, one shell pass for the condensing steam, and an even number of tube passes for the sulfuric acid solution.

The evaporator design must address the peculiar physical properties of concentrated sulfuric acid to 70 wt% H_2SO_4, especially the reduction of operating pressure and the increase of operating temperature due to the low sulfuric acid vapor pressure; the installation of a demister tower to reduce the entrainment of sulfuric acid mist with the water vapor.

12.1.11.3 Triple-Effect Calculations

Let's consider three evaporators arranged in series connected to form a triple-effect system as depicted in Figure 41. Connections are made in order that the vapor generated from one effect serves as the heating steam for the next step.

A condenser and a vacuum ejector establish a vacuum or reduced pressure in the final or third effect and eventually withdraw non condensable gases from the final concentrated sulfuric acid product.

The first effect as depicted is the first stage to which the feed stream of dilute sulfuric acid is heated directly by injecting raw steam inside the heat exchanger and in which the pressure in the vapor space is the highest. In this manner, the pressure differential between the steam and the condenser is spread across the remaining two effects.

The pressures p_1, p_2, and p_3, in each effect is lower than that in the preceding effect and higher than the subsequent step: $p_1 > p_2 > p_3$

When steady operation is reached, the mass flow rates are such that neither water nor sulfuric acid accumulates or depletes in any of the effects.

The temperature, concentration and flow rate of the feed are fixed, the pressure in steam inlet and condenser established and all liquor levels in each effect maintained.

Then all internal concentrations, flow rates, pressure and temperature are automatically kept constant by the operation. Hence, the concentration of the final product can be change only by changing the feeding rate.

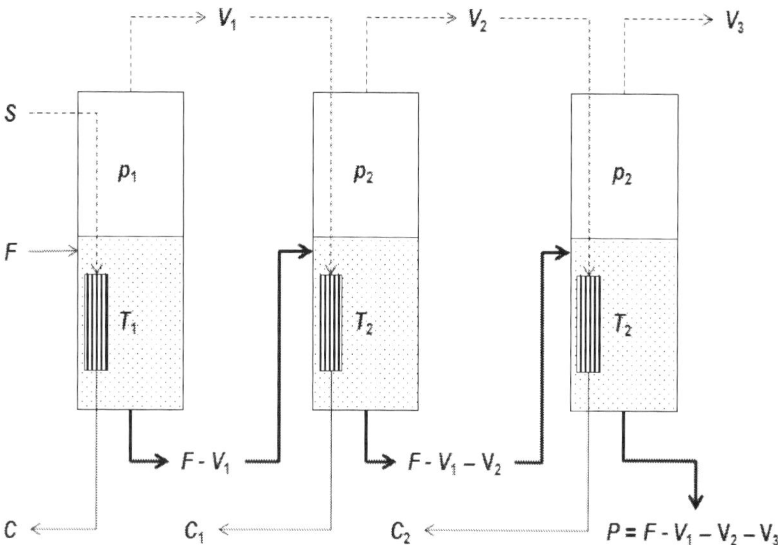

Figure 41 – Schematic of a triple-effect evaporator design with forward feed

Therefore, the heat flow rate across each effect can be described as follows if we neglect in a first approximation the boiling point elevation (BPE) in each effect and considering that the heat flow rate in the first effect serves only for vaporizing the feed (i.e., latent enthalpy of vaporization) and that the sensible heat required to pre-heat the feed to the boiling point is neglected, we have the following equations with the temperature of the condensate leaving one effect being the temperature of boiling for the next effect:

Effect No. 1 $\quad Q_1 = U_1 A_1 \Delta T_1 \quad$ with $\quad \Delta T_1 = T_1 - T_F$

Effect No. 2 $\quad Q_2 = U_2 A_2 \Delta T_2 \quad$ with $\quad \Delta T_2 = T_1 - T_2$

Effect No. 3 $\quad Q_3 = U_3 A_3 \Delta T_3 \quad$ with $\quad \Delta T_3 = T_2 - T_3$

Therefore, the overall temperature difference, denoted ΔT in K, equals simply the sum of the temperature difference in each effect:

$$\Delta T = \Delta T_1 + \Delta T_2 + \Delta T_3$$

On the other hand, the total capacity is proportional to the overall heat flow rate, denoted Q_T, equals to the sum of each single contribution:

$$Q_T = UA\Delta T = Q_1 + Q_2 + Q_3 = U_1 A_1 \Delta T_1 + U_2 A_2 \Delta T_2 + U_3 A_3 \Delta T_3$$

In practice, in order to obtain an economy of construction, all the heating surface areas in the three effects of the triple-effect system are kept equals:

$$A = A_1 = A_2 = A_3$$

On the other hand, the temperature of the condensate from the first effect equals that of the boiling point of the feed of the second effect, and the temperature of the condensate of the second effect equals the boiling point of the feed of the last effect, we can assume that all the heat flows are also equals:

$$Q_T = Q_1 = Q_2 = Q_3$$

Therefore, in a first approximation, the overall heat transfer coefficient in the three-effect evaporator is inversely proportional to the reciprocal of each single heat transfer coefficient:

$$U = 1/[1/U_1 + 1/U_2 + 1/U_3]$$

Hence, the temperature difference for each effect is simply given by the following three equations:

Effect No. 1 $\qquad \Delta T_1 = [\Delta T/U_1]/[1/U_1 + 1/U_2 + 1/U_3]$

Effect No. 2 $\qquad \Delta T_2 = [\Delta T/U_2]/[1/U_1 + 1/U_2 + 1/U_3]$

Effect No. 31 $\qquad \Delta T_3 = [\Delta T/U_3]/[1/U_1 + 1/U_2 + 1/U_3]$

12.1.11.4 Forward Feed Calculations

If the vapor space of the third effect undergoes a reduced pressure of 200 mbar (2 kPa), then the sulfuric acid product (70 wt.% H_2SO_4) will boils at 100°C (212°F). Moreover, if we use as before saturated steam at 1.805 MPa (262 psia) with a condensing temperature of 207°C (405°F), the overall temperature differential across the evaporator is 107°C (245°F)

Therefore, if we assume that the heat transfer coefficients for the first, second, and third effect are in decreasing order 4,000 W/m²/K, 3,000 W/m²/K, and 1,500 W/m²/K respectively because the viscosity of each product increases, the three temperature differences are hence calculated using the previously established equations as 21°C, 29°C, and 57°C.

12.1.11.5 Concentration of Sulfuric Acid from 70% to 93.3%

The main drawback of the triple-effect forward-feed evaporator is that the most concentrated sulfuric acid is produced in the third effect, where the temperature is the lowest. Therefore because the dynamic viscosity of the concentrated sulfuric acid reaches its maximum the overall heat transfer coefficient is smaller compared to the one in the first and second effects. This is the main reason why it is preferable to run the triple-effect system up to a maximum strength of sulfuric acid of 70 wt.% H_2SO_4.

Therefore the final concentration stage is performed using a horizontal boiler. This type of horizontal boiler for the final concentration of the sulfuric acid is commercialized by the leading manufacturer *QVF-DeDietrich Process Systems GmbH*, Mainz, Germany.

The major advantages of such approach according to the manufacturer are as follows: (i) it allows the best use of the tantalum heating surfaces due to the concentration profile throughout the length of the evaporator with a high overall heat transfer coefficient reaching 4,000 $W/m^2/K$; (ii) low fluid pressure exerted on the heating surfaces, thus low-pressure differentials inside the evaporator; (iii) minimizing the sulfuric acid losses in the exhaust steam; (iv) small sulfuric acid volumes immobilized inside the evaporator during operation ensuring a safe and flexible operation; and finally (v) no mechanical moving parts (i.e., no pumps).

12.1.12 Mechanical Vapor Recompression (MVR)

In addition to the possibility of recovering the steam from one effect to the next, a further improvement for steam economy, is to take the vapor from the first effect and, after compressing it, return it to the steam chest of the evaporator from which it was evaporated. The compression can be performed either by using some fresh steam, at a suitably high pressure, in a jet ejector pump, or simply by using a mechanical compressor. However, the utilization of jet ejectors is the more common mean nowadays. Therefore, a proportion of the vapors are re-used, together with fresh steam, and hence a considerable overall steam economy is achieved by reusing the latent heat of vaporization over again.

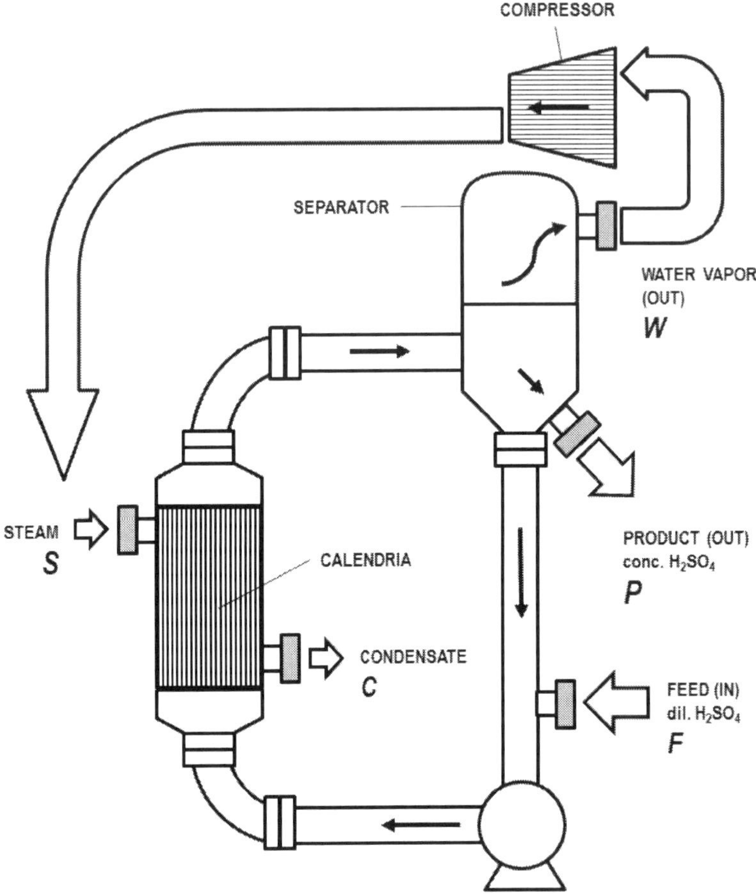

Figure 42 – Evaporator with mechanical vapor recompression

12.2 Concentration by Acid Retardation (APU)

The regeneration of spent sulfuric acid and other mixtures of sulfuric acid and its salts can be performed by the so-called **acid retardation separation** or. In this process, the spent sulfuric acid solution containing high concentration of metal sulfates circulates through a column packed with an ion exchange resin of the type Dowex MSA-1 (Trademark of the *The Dow Chemical company*) [176] that specifically absorbs the acidity from the solution.

[176] Collective (2001) *Dowex Ion Exchange Resins. Powerful Chemical Processing Tools*. Form No. 177-01395-602QRP, The Dow Chemical Company, Houston, TX.

The absorption of the strong sulfuric acid is favored by the selected ion exchange resin; hence the displacement of the acid through the resin bed will be retarded relative to the movement of the metal cations as in other chromatographic separation. Once all the resin bed has be loaded and the breakthrough of acid is detected, the regeneration of the resin is performed by eluting it with water passing through the resin bed counter currently. This generates a concentrated sulfuric acid solution and weak acid by-product high metal content. For more information, we [177]

The process is commercialized under the acronym APU™ by the Canadian company *ECO-TEC* (Pickering, ON) that was acquired by *Koch Separation Solutions* (Wilmington, MA) and rebranded AnoPur™.

---

[177] BROWN, C.J. (1999) *Mixed acid recovery with the APU acid sorption system. An update.* Technical paper 147, Eco-Tec, Pickering, ON, Canada.

13 Sulfation Economics

13.1 Capital Expenditure (CAPEX)

Below are reported some order of magnitude for the fixed costs (FC), the working capital (WC), and capital expenditure (CAPEX) for selected sulfation processes based on techno-economic reports, preliminary economic assessment (PEA) or preliminary feasibility study reports, and news releases.

Table 85 – CAPEX of selected sulfation plants

Sulfation process	Nameplate capacity of conc. (tonnes/year)	Product (tonnes/year)	Fixed costs (million dollars)	Working capital (million dollars)	CAPEX (million dollars)	Capital intensity (CAPEX/tonne product)
Aluminum sulfate plant	50,000 bauxite (65 wt.% Al_2O_3)	109,000 tonnes $Al_2(SO_4)_3$	$95	$10	$105	$963
Phosphate plant	4,000,000 phosphate rock (70% BPL)	1.34 million tonnes P_2O_5	$175	$25	$200	$182
Titanium sulfate plant	200,000 ilmenite (55 wt.% TiO_2)	100,000 tonnes TiO_2	$265	$46	$311	$3,110
Lithium plant	217,000 spodumene (5.5 wt.% Li_2O)	9,765 tonnes Li_2CO_3	$213	$31	$244	$24,987
Rare earth plant (*)	55,000 monazite (33 wt.% TREO)	18,000 tonnes TREO (3,500 tonnes NdPr)	$400	$60	$460	$26,000
Notes: (*) Acid-cracking, leaching, and separation plant; (**) Costs in US dollar (USD)						

These data must be used with great caution as they originate from various sources and must be use as rough indicators.

13.2 Operating Expenses (OPEX)

Below are reported some order of magnitude for the operating cost (OPEX) and manpower for selected sulfation processes based on techno-economic reports. As previously, these data must be used with great caution as they originate from various sources and must be use as rough indicators.

Table 86 – OPEX of various industrial sulfation processes

Sulfation process	Nameplate capacity (tonnes/year) of concentrate	OPEX ($/tonne of conc.)	Direct labour (man-hour/tonne)
Aluminum sulfate plant	50,000 bauxite (65 wt.% Al_2O_3)	$100	1.5
Phosphate plant	4,000,000 Phosphate rock (70% BPL)	$66	0.1
Titanium dioxide plant	200,000 ilmenite (55 wt.% TiO_2)	$2,030	0.5
Lithium plant	217,000 spodumene (5 wt.% Li_2O)	$131	1.0
Rare earths(*)	55,000 monazite (33 wt.% TREO)	$3,000	2.0
Notes:(*) Acid-cracking, leaching and separation by solvent extraction (SX) ; (**) Costs in US dollar (USD)			

14 Prototype and Pilot Testing

The prototype and pilot testing of the sulfuric acid digestion, the sulfation baking, and the sulfation roasting require out-of-shelves and custom made acid resistant vessels or digesters, reactors or rotary kilns capable to resist both corrosion under harsh operating conditions (i.e., strong acidity, relatively high temperature) together with fast heating rates related to strong exothermic sulfation reactions and thus are always prone to thermal runaway if not properly controlled.

Worldwide, a handful of industrial and corporate laboratories possess such technical capabilities. All these equipment were always devised and constructed in-house and are mostly used in the titanium pigment industry, the phosphate industry, and in a lesser extent in the lithium industry.

The present section provide some key experimental insight from the prototype and semi-pilot testing installations used to process a varieties of mineral concentrates such as phosphate rock, magnetite, ilmenite, slags, bauxite, and bauxite residues, phosphate rock, pyrochlore, and spodumene to name a few that were processed and tested during the last thirteen years by *Electrochem Technologies & Materials Inc.* at the company facilities located in Boucherville, Quebec, Canada.

14.1 Prototype Testing

14.1.1 Prototype Setup

At Electrochem, the prototype tests are all conducted inside a one-gallon digester vessel with a brim capacity of 4.650 L that consists to a modified tall cylinder made of PFA with 6¼-inch (158.75 mm) inner diameter and 9¼-inch (23.495 mm) with 1/8-inch (3.173 mm) thick walls from the US company *Savillex*, (Eden Prairie, MN).

The vessel was equipped with a lid and a tightening ring made of ECTFE reinforced by a ring of chemically pure titanium (ASTM grade 2). In addition to the two existing PFA fittings installed by the manufacturer including a safety vent, two additional PFA compression fittings were also installed: (1) a bore through fitting allowing a PTFE-lined Type-K thermocouple probe to be immersed into the mixture, and (2) an exhaust 3/8-inch (9.525 mm) PFA tube to evacuate steam, and fumes. The system can be used safely up to the maximum temperature of 260°C (500°F).

The digester is usually heated inside a 5.5 kW electric kiln. Actually, a minimum external heating is required to bring the temperature up to the water injection temperature if needed and also during the sulfation baking period to compensate for the heat losses as the bare vessel is not insulated.

The exhaust tube is connected to a tall absorber column packed with ½-inch OD (12.7 mm) polypropylene balls, and scrubbed by recirculating alkaline water using a diaphragm pump with a volume flow rate of 3 gallons (US) per minutes (11.36 L/min) in order to absorb rapidly all the noxious SO_2 fumes, the entrained sulfuric acid mist, and also to condense the steam.

In case, the sulfation roasting is required an horizontal kiln equipped with a 2-inch OD (5.08 cm) tube made of fused quartz or electrofused alumina can be installed instead and the mixture of sulfuric acid and solid feed with a pug mill or a screw feeder.

The step-by-step description for performing the sulfation at the prototype scale are summarized and depicted in Figure 43.

14.1.2 Protocol for Injecting Water

If needed, it is possible to perform the injection of water at a temperature around 80-95°C at which a fixed amount of water is introduced rapidly by means of a peristaltic or syringe pump. The intent of this ancillary technique is to raise the temperature of the mixture by simply releasing the heat of hydration of the sulfuric acid.

The amount of water (m_{H2O}) required based on the mass (m_A) and initial concentration of sulfuric acid (w_A) and the final targeted concentration during the digestion (w_A*) is simply given by the equation derived from the mixing rules (see Section 18.3):

$$m_{H2O} = m_A \, (w_A - w_A*)/w_A*$$

14.1.3 Baking

Afterwards, the setting of the mixture into a hard sulfation cake is allowed to cure at a **_baking temperature_** for several hours.

For the entire duration of the prototype testing, it is of paramount importance to record accurately the temperature every second using a data logger using several type-K thermocouples immersed directly inside the slurry with the measuring tip protected by a custom-made corrosion resistant sleeve made of PTFE or tantalum sheaths.

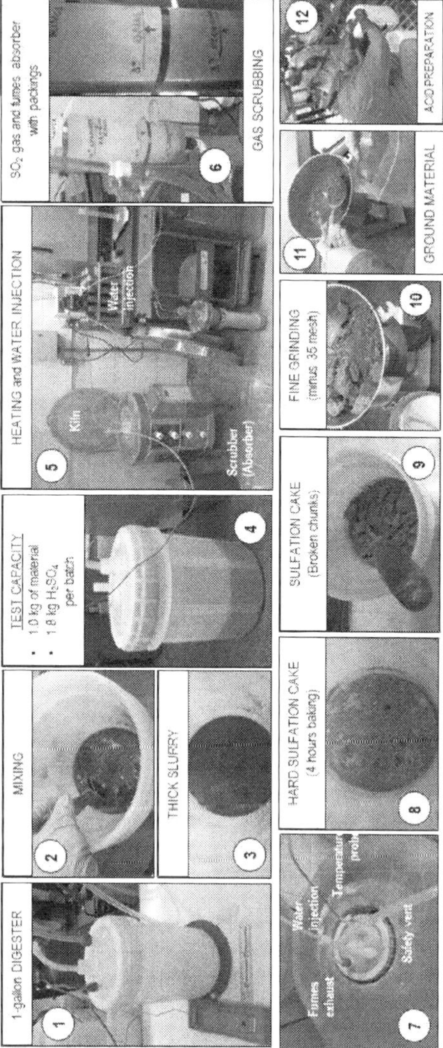

Figure 43 – Electrochem's one-gallon PFA digester setup and protocol

Usually, each plot of the temperature vs. time allows to identify clearly the heating region, the steep increase of the temperature during the injection of water and set-off that triggers the exothermic sulfation reaction always reaching a peak temperature that strongly depend on the material to be sulfated, the acid-to-solid ratio the concentration of acid at attack, etc.

An experimental temperature plot showing a typical "surge wave" with a S-shape recorded during the sulfation of vanadiferous titano-magnetite (VTM) performed at the prototype scale by the author is depicted in Figure 44.

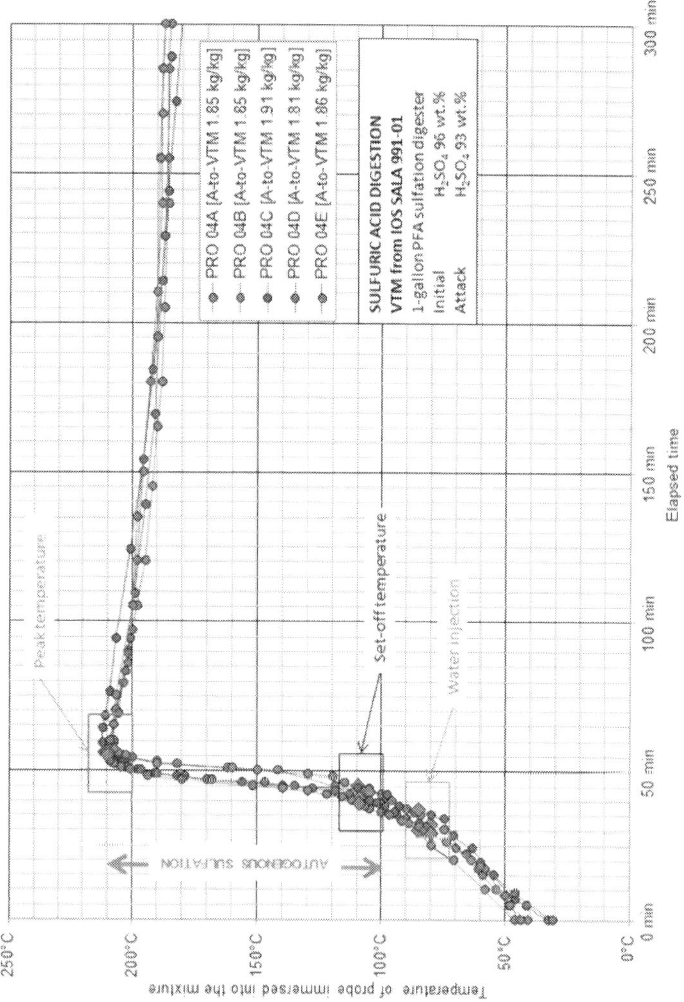

Figure 44 – Plot of temperature vs. time for a typical sulfation tests

14.2 Pilot Testing

14.2.1 Construction of 15-gallon Digester

When the experimental results obtained during the trials using the prototype digester testing needs to be validated at a semi-pilot scale, a custom made digester with 15-gallon (ca. 56.8 liters) brim capacity was designed and built to mimic the industrial scale digesters.

An illustration of a schematic drawing illustrating a custom made 15-gallon sulfation digester used for preforming the sulfuric acid digestion of various vanadiferous feedstocks is depicted in Figure 45.

The digester body consists of a large 12-inch (30.48 cm) NPS × 8-inch (20.32 cm) reducing tee (03) made of a steel shell (Schedule No. 40), lined internally with a ¼-inch thick PTFE liner (04) manufactured by *Crane Resistoflex* (Marion, NC, USA) with ANSI Class 150 ductile cast iron bottom flanges (01) and a top lower flange (06).

14.2.2 Pilot Digester Installation Setup

The PTFE thick lining mimics to some extent the brick lining typical of industrial scale digesters. The reducing tee was installed in the upright position onto a 2000-lb (907.18 kg) heavy duty cart (21) with the 12-in (30.48 cm) bottom flange sitting onto the skid frame while a 12-in top (30.48 cm) stainless steel flange (07) with an auxiliary 6-in (15.24 cm) flange (09) were constructed in order to accommodate several ports and measuring devices as follows: (a) a ¾-inch (1.905 cm) deep pipe connected to a gas line (15) and a water injection line (17) by means of a three-way valve (16); (b) a second ⅜-inch (9.525 mm) deep inlet pipe (10) for introducing the sulfuric acid feed connected a diaphragm pump; (c) a ¾-inch exhaust pipe (13) with a ball valve (14); (d) a pressure relieve toggle valve (11) and pressure gauge (19); (e) three Type K thermocouple probes (8, 12, and 18) with a protective PFA sheath; and (f) an auxiliary 6-inch (15.24 cm) NPS flange (09) for introducing the 4-inch (10.16 cm) diameter impeller shaft.

The 8-inch (20.32 cm) side flange (20) was only used as manhole for accessing the inside of the vessel for maintenance and/or cleaning. Moreover, two 4-inch (10.16 cm) wide high temperature silicone band heaters (2 and 5) were installed on the digester's outer shell in order to supply a minimum amount of heat while a 2-inch (5.08 cm) thick thermal insulation made of **Kaowool®** (Trademark of *Thermal Ceramics Inc.*) was added to minimize heat losses during autogenous operation. A photograph illustrating the custom made 15-gallon sulfation digester is depicted in Figure 45.

The schematic process flow diagram (PFD) illustrating the custom made 15-gallon sulfation digester is depicted in Figure 46.

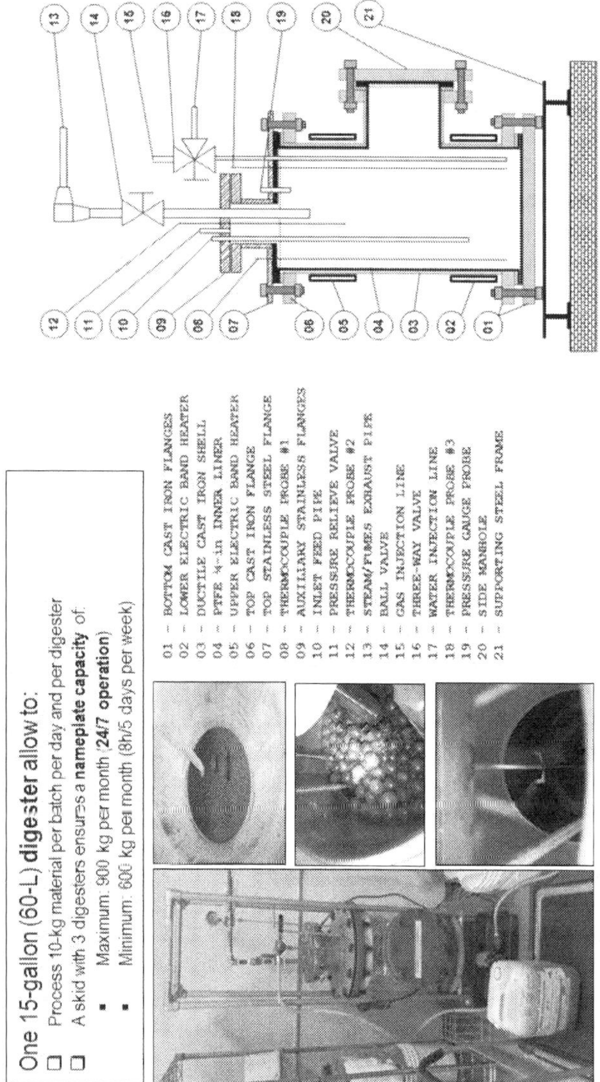

Figure 45 – Electrochem's 15-gallon PTFE-lined pilot digester

14.2.3 Mode of Operation

The concentrated sulfuric acid is introduced into the 15-gallon digester by means of a diaphragm pump. In order to prevent blockage of the immersed tip of the water injection line, compressed air or inert gas is introduced through a three-way valve until the requirement for water injection into the system.

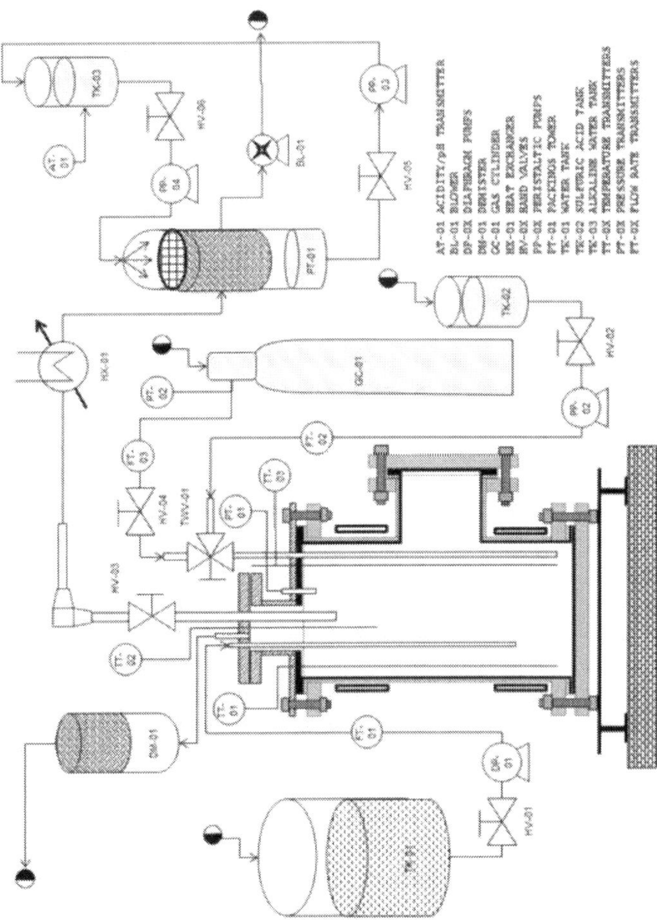

Figure 46 – Pilot digester installation

The predetermined weighted amount of water, contained in a small water tank, is injected into the mixture using a peristaltic pump or by a simple displacement method using compressed air as vector.

Eventually, any excessive pressure build-up is released by means of a toggle relieve valve connected to a demister. The fumes and steam produced during the digestion are extracted through the exhaust pipe using a blower and are scrubbed by spraying an alkaline solution over a tall column packed with plastic internals. A coiled PFA heat exchanger was also used to lower the temperature and to condense the steam.

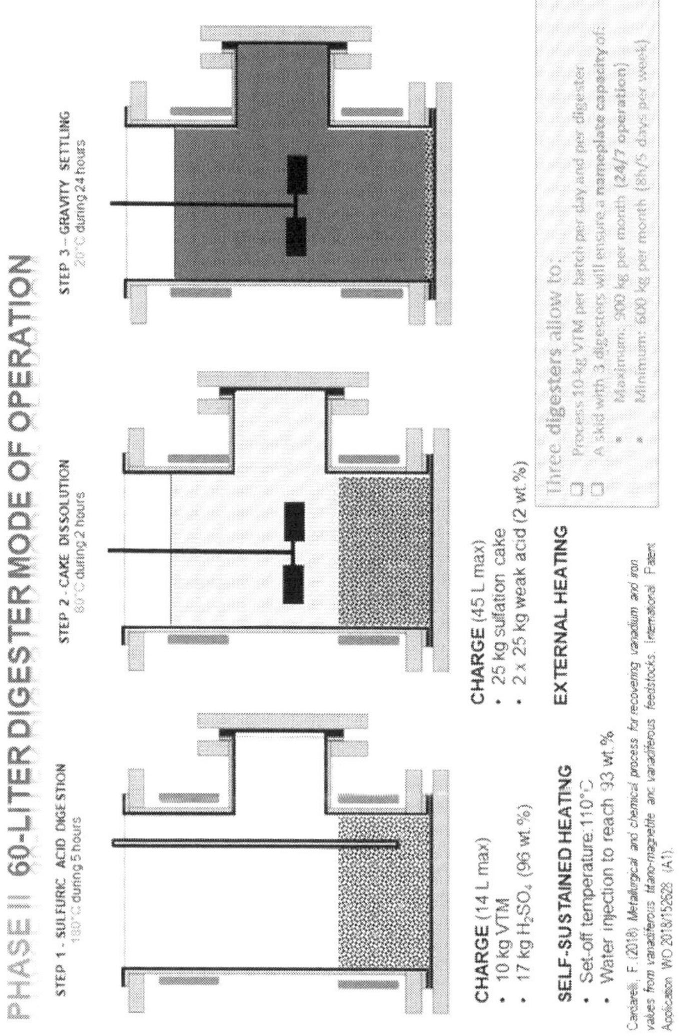

Figure 47 – Mode of operation of 15-gallon pilot digester

The installation depicted is capable of processing 10 kg (22.05 lb.) of solid feedstock with up to 20 kg (44.09 lb.) of concentrated sulfuric acid per batch that corresponds to a charge occupying only 25 vol.% of the brim capacity (ca. 14 L).

Once the sulfuric acid digestion is completed (Step 1) and the temperature inside the digester decreased down to 80° C (following several hours of digestion), the digester is filled with up to 45 kg (99.2 lb.) of warm acidified water

Afterwards, a 4-inch (10.16 mm) impeller is introduced and operated with a 600 rev/min (62.8 rad/s) motor. Such in-situ dissolution of the sulfation cake is usually completed within one to two hours of intense mixing (Step 2).

Finally, the agitation is ceased and followed by gravity settling of the solids; only then the digester is emptied by pumping out the supernatant pregnant solution.

15 Health and Safety

Concentrated sulfuric acid is an extremely corrosive and hygroscopic substance, but it can be handled safely if the proper precautionary measures are implemented by qualified safety professional and strictly observed. This section only presents some peculiar measures regarding the handling, the dilution with water, and the cleaning of tanks and vessels.

For detailed information regarding standard Environmental, Health & Safety (EH&S) measures we recommend the reader to consult sulfuric acid producers and suppliers reference handbooks [178, 179]

15.1 Corrosiveness and Handling

The handling of aqueous solutions of sulfuric acid with a concentration below 40 wt.% H_2SO_4 are no riskier than the manipulation of aqueous solutions of hydrochloric and nitric acids with similar strength and thus they do not represent a significant additional risk and they are relatively safe to work with provision similar safety procedures are in place and chemical operators wear their usual protective personal equipment (PPI) such as acid-proof coveralls, gloves, safety rubber boots, goggles, face shields, hard hats, and respirator.

However, concentrated sulfuric acid solutions with a mass percentage greater than 80 wt.% H_2SO_4 becomes extremely corrosive and dangerous from an occupational health and safety standpoint. It does not only cause chemical burns, it also causes burns by dehydration of organic materials (i.e., caustic) including skin and animal tissues, destroying entirely the compounds. This is exemplified by the well-known charring of carbohydrates such as sugars and cellulosic materials that sulfuric acid breaks down to water vapor and a voluminous expanded charcoal mass in a matter of seconds.

Moreover, when heating concentrated sulfuric acid, it is important to recall that the liquid exhibits a volume expansion of nearly 15 vol.% between ambient temperature and 330°C. Thus it is advised to take suitable measures to prevent overflow and spills.

---

[178] COLLECTIVE (2007) *Sulfuric Acid Handbook. An information source for industrial consumers, handlers, transporters and other users.* Norfalco LLC., Independence, OH.

[179] COLLECTIVE (2002) *Sulfuric Acid Chemical Safety Handbook.* Southern States Chemicals, Dulany Industries Incorporated, Savannah, GA.

Finally, hot concentrated sulfuric acid when reaching the boiling start to decompose to yield copious white fumes made of tiny droplets of sulfuric acid, mixed with sulfur dioxide and traces of sulfur trioxide, forming highly toxic and corrosive aerosols that are classified as lung carcinogen.

15.2 Controlled Dilution of Sulfuric Acid

When diluting concentrated sulfuric acid with water because of the large hydration enthalpy already discussed in Section 8.7 the process is highly exothermic releasing large quantities of heat that can result in the boiling of the mixture if the process is not carefully controlled.

Thus the continuous production of sulfuric acid at a given concentration through inline dilution of concentrated sulfuric acid must be addressed very carefully with dedicated equipment to ensure the safety of the workers.

In practice, the dilution to reach the designated acid strength is usually performed in specially designed equipment including an inline static dispersion mixer, followed by fast cooling by means of a plate and frame heat exchanger such as those manufactured by the US Company *InyoProcess* and the Japanese company *Noritake*.

Example: The two stages dilution method for preparing 1,920 kg of an aqueous solution of 50 wt.% H_2SO_4 at 25°C (77°F) from 1,000 kg concentrated acid with a strength of 96 wt.% H_2SO_4 and 920 kg of deionized water both at an initial temperature of 23.9°C (75°F) are summarized below and also depicted in Figure 48 showing the masses involved, and the amount of heat released for each stage.

During the first stage, the exact amount of water required to obtain 80 wt.% sulfuric acid are simply calculated using the mixing rules: [(96 − 80)/(80 − 0)] x 1,000 kg = 200 kg. Based on the specific enthalpies of 96 wt.% sulfuric acid at 23.9°C (-58.9 kJ/kg), 80 wt.% sulfuric acid at 37.8°C (-194.1 kJ/kg), and water at 23.9°C (+100 kJ/kg), the maximum (adiabatic) release of heat is 194.0 MJ.

During the second stage, the exact amount of water required to obtain 50 wt.% sulfuric acid are again calculated using the mixing rules: [(80 − 50)/(50 − 0)] x 1,200 kg = 720 kg. Based on the specific enthalpies of 80 wt.% sulfuric acid at 37.8°C (-194.1 kJ/kg), 50 wt.% sulfuric acid at 25°C (-220.3 kJ/kg), and water at 23.9°C (100 kJ/kg), the maximum (adiabatic) release of heat is 90.3 MJ. It is interesting to note that the latter amount represents only 32% of the total heat released (284.3 MJ).

This type of stepwise approach emphasizes the benefits of performing the dilution in at least two consecutive steps to prevent overheating and to avoid a thermal runaway.

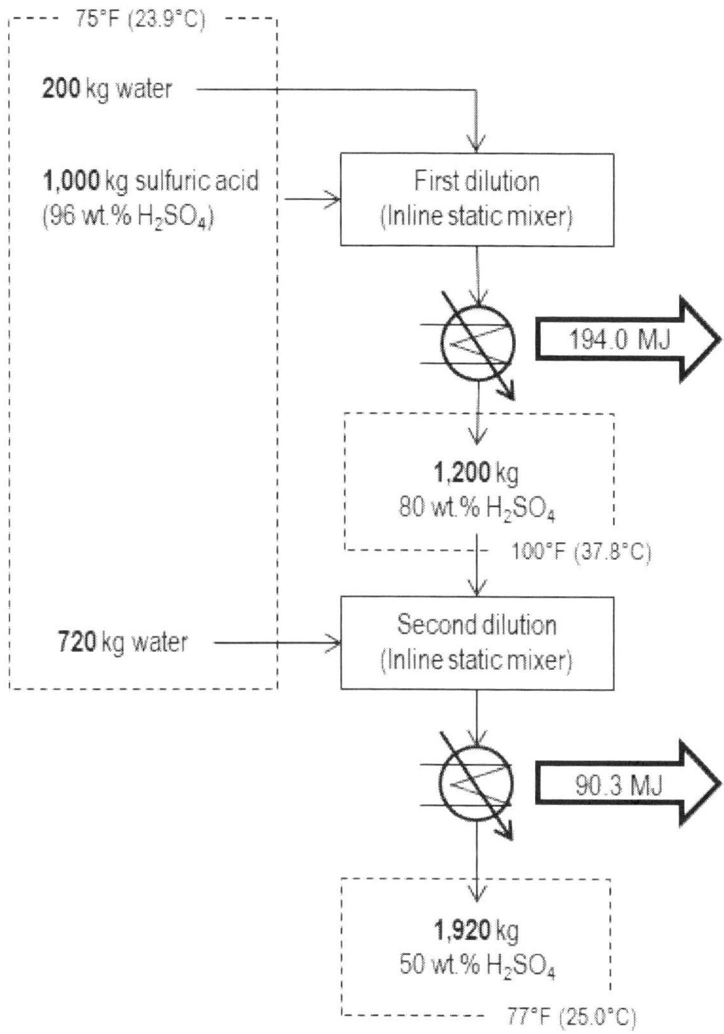

Figure 48 – Flow diagram for the two-step dilution of concentrated sulfuric acid

15.3 Cleaning and Flushing Vessels, and Lines

Because oleums react explosively with water, never flush out oleum or concentrated sulfuric acid from digesters or reactors directly with water as it will result in catastrophic thermal runaways. Always start the cleaning procedure incrementally with a sulfuric acid strength which is weaker than the previously filled one. Follow the sequence, given in Figure 49, before cleaning with water.

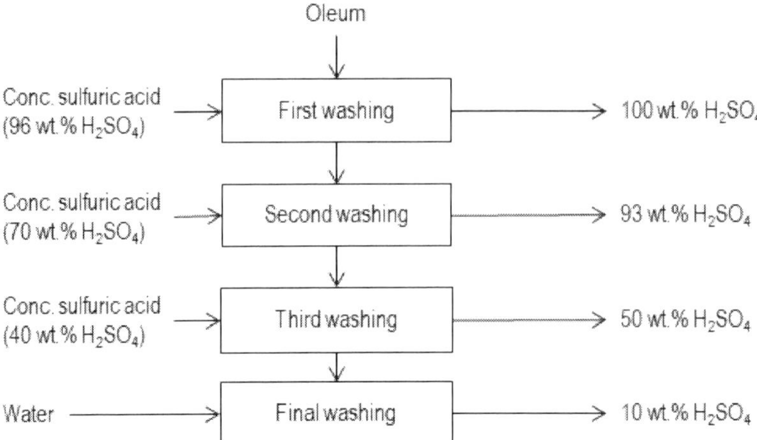

Figure 49 – Cleaning protocol for vessels containing sulfuric acid and oleum

16 Analysis

16.1 Densitometry, Sound Velocity, and Refractometry

Regarding the determination of the mass percentage of sulfuric acid by density, sound velocity, and refractometry, these techniques were already discussed in Sections 5.1.2, 5.3, and 5.9.4 respectively.

16.2 Acid-Base Titration

The conventional method for determination of sulfuric acid concentration in the absence of interfering metallic cations is acid-base titration. This determination of free sulfuric acid in an aqueous solution is a straightforward analytical protocol as it can be performed by common acid-base titration of the first acidity with a pH indicator such methyl orange or better by potentiometry using a pH indicating glass electrode for the titration and detecting the end point by mathematical treatment of the data. In routine analysis accuracies of 0.1 wt.% to 0.5 wt.% H_2SO_4 can be attained.

However, in case of highly concentrated solutions or even oleums, the sampling followed by proper dilution is tedious and even hazardous for the latter. Moreover, in the particular case of spent solutions or pregnant leach solutions those are dark and/or strongly colored solutions that always contains significant amount of amphoteric metallic cations (e.g., Fe, Cr, Ni, Mn, V) that hydrolyze at acidic pH and precipitate as metal hydroxides well before the neutral point is reached. Thus the abnormal consumption of base for precipitating the metal hydroxides prevents the accurate determination of the free sulfuric acid. If only ferrous sulfate is present in significant amount theoretically it could be possible to perform the acid-base titration by means of sodium carbonate using methyl orange as indicator or to add a complexing agent such as sodium fluoride or sodium EDTA preventing the Fe(III) to interfere. Because most of the time, ferric iron is also present the precipitation of ferric hydroxide limits this method to fully reduced liquors not exposed to air.

On the other hand, because of the high acidity, the pH cannot be measured accurately with glass electrodes. Even an ancillary hydrogen gas reference electrode yields experimentally unreliable results because of the presence of interfering reducible cations and poisoning impurities (e.g, carbon and silica colloidal particles). Moreover, at such high concentration of protons the calculation of the concentration of a diprotic acid such as the sulfuric acid is complicated as it requires knowing accurately the activity coefficients and dissociation constants for the second acidity.

16.3 Gas Volumetric Analysis

Gas volumetric analysis based on the utilisation of a gas burette for the quantitative determination of various inorganic chemicals that evolve gases upon reacting with certain substances such as carbonates, oxalic acid, hypochlorite and hydrogen peroxide including free-sulfuric acid by means of the gas burette.

Actually, the set-up is an improved and modified version of the original Bernard's or ***Dietrich-Frühling*** calcimeter used extensively by the author in the 1980s for the determination of calcium carbonates in rocks and soils (see Figure 50). The setup consists of a graduated glass cylinder with two stopcocks at each end, also called a gas burette, which is connected at the bottom to a holding glass reservoir by flexible plastic tubing made of Tygon® (Trademark of *Saint-Gobain Performance Plastics Corporation*), while the top of the burette is connected to the reaction Erlenmeyer flask with a ground joint tap. Both the burette and the reservoir can be moved vertically along a tall vertical stand in order to adjust and equilibrate the liquid column inside.

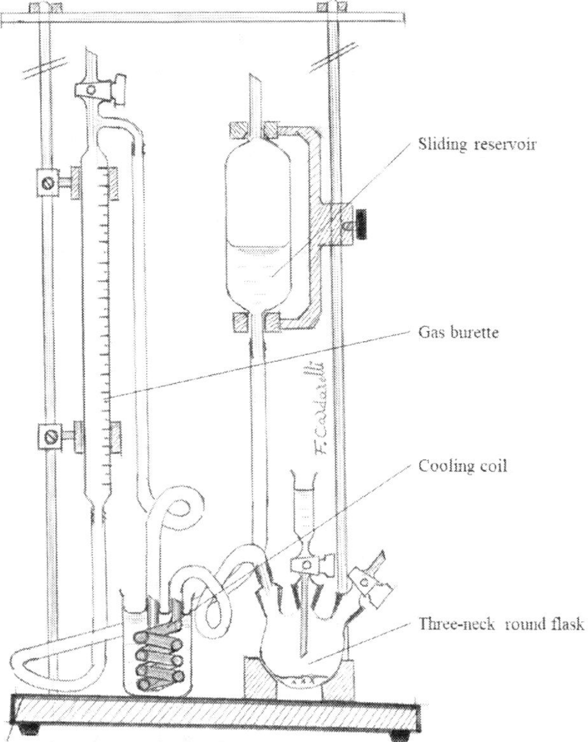

Figure 50 – Dietrich-Frühling gas burette apparatus

The determination is based on the chemical reaction used to generate the carbon dioxide gas that consists simply to the chemical reaction of an alkali metal carbonate or alkali-earth metal carbonate in an acidic solution as follows:

$$Na_2CO_3(s) + 2H^+ = 2Na^+ + CO_2(g)$$

$$K_2CO_3(s) + 2H^+ = 2K^+ + CO_2(g)$$

$$Cs_2CO_3(s) + 2H^+ = 2Cs^+ + CO_2(g)$$

$$CaCO_3(s) + 2H^+ = Ca^{2+} + CO_2(g)$$

$$MgCO_3(s) + 2H^+ = Mg^{2+} + CO_2(g)$$

Among them potassium and cesium carbonates are preferred over sodium carbonate as they form a highly concentrated lye without crystallization during temperature changes because the resulting potassium and cesium sulfates are highly water soluble.

Similarly, it is also possible to use zinc or iron metal powders instead of alkali-metal carbonates that readily evolve nascent hydrogen gas upon contact with the free sulfuric acid according to:

$$Fe(s) + 2H^+ = Fe^{2+} + H_2(g)$$

$$Zn(s) + 2H^+ = Zn^{2+} + H_2(g)$$

However, we observed in practice that the release of hydrogen gas is not as fast as with the carbon dioxide in the case of carbonates and often mild external heating of the round flask is required to speed up the dissolution reaction affecting the volume measurements.

It is important that the gas burette and the mobile reservoir are filled with the **_Kobe's solution_** which is a concentrated aqueous solution of sodium sulfate (Na_2SO_4 20 wt.%) with 5 wt.% of free-sulfuric acid prepared from freshly boiled deionized water and afterwards saturated with carbon dioxide gas prior to perform measurements [180, 181].

Actually, a **_modified Kobe's solution_** using potassium sulfate instead of the sodium sulfate version offers better measurements as the latter has a tendency to crystalize inside the burette and tubing causing blockages.

Such solution is used as the solubilities of several gases (oxygen, hydrogen, and nitrogen) including carbon dioxide gas at 20°C are significantly reduced as indicated by the lower Bunsen coefficient (expressed in cm^3 of pure gas per liter of solution) [182] when compared to pure water at the same temperature (Table 87).

[180] KOBE, K.A.; and WILLIAMS, J.S. (1935) Confining liquids for gas analysis: solubility of carbon dioxide in salt solutions. *Industrial & Engineering Chemistry Analytical Edition* **7**(1)37-38.

[181] KOBE, K.A.; and KENTON, F.H. (1938) Confining liquids for gas analysis: solubility of common gases in acid sodium sulfate solution. *Industrial & Engineering Chemistry Analytical Edition* **10**(2)76-77.

[182] GUERIN, H. (1981) *Traité de manipulation et d'analyse des gaz, seconde édition.* Masson & Cie, Paris, France.

Table 87 – Solubility of H_2 and CO_2 in water and Kolbe solutions

Gas	Pure water (0°C)	Pure water (20°C)	Kolbe solution (Na_2SO_4) (25°C)	Kolbe solution (K_2SO_4) (25°C)
Hydrogen (H_2)	0.021	0.018	0.007	n.m.
Nitrogen (O_2)	0.023	0.016	0.005	n.m.
Oxygen (O_2)	0.049	0.031	0.009	n.m.
Carbon dioxide (CO_2)	1.795	0.874	0.27	0.20

This precaution is mandatory in order to reduce the error due to the dissolution of carbon dioxide gas and/or the simultaneous release of dissolved oxygen and nitrogen gases.

Historically, liquid mercury definitely offered a better measurement accuracy as carbon dioxide gas dissolution becomes irrelevant but due to its elevate toxicity and the need to use thicker and more mechanically robust and expensive glassware to handle the denser liquid mercury, it is usually recommended to use the above aqueous solution instead.

The protocol consists: (1) to pour an excess amount of saturated solution of potassium carbonate inside the narrow 10-mL separatory funnel with stopcock closed and connected through a pierced cork and the tubing to the gas burette)not shown on the drawing), then (2) a precise aliquot of the solution to be analyzed which is weighted accurately is introduced inside the 100-mL round flask. The rubber cork and the connection tubing are immediately inserted firmly to reduce the losses. (3) Once closed, the level of the liquid in the gas burette is adjusted by sliding the reservoir vertically until the zeroing is completed. (4) Then, the stopcock is open and release immediately the lye of potassium carbonate that reacts immediately with the free acid and effervesces steadily evolving carbon dioxide

Alternatively, the saturated solution of potassium carbonate can be contained inside a small test tube and introduced into an Erlenmeyer flask. Only once the setup is closed and the level of liquid is equilibrated the Erlenmeyer is inclined to allow the liquid to spill and react with the free acid from the sample.

Once the reaction is almost complete, that is, the gas evolution has ceased, the flask is immersed into a water bath at room temperature in order to stabilize its temperature before the readings. Another option is to use a cooling tin coil immersed in water at room temperature.

The level of the liquid in the gas burette and reservoir are made equal by displacing the gas burette and reservoir vertically and the volume of carbon dioxide gas measured, together with the ambient temperature and pressure conditions in order to calculate its normal volume (NTP). Usually, duplicate or triplicate measures are required to get the most reliable result.

The volume of carbon dioxide gas under normal temperature and pressure conditions (NTP), that is, $T_0 = 273.15K$ and $P_0 = 101.325$ kPa is simply obtained from **Gay-Lussac** and **Charles laws** and correcting the total carbon dioxide gas pressure subtracting the vapor pressure of water, π_V, at the measured ambient temperature as follows:

$$V_{CO2}(NTP) = V_{CO2}(T_0/T)[(P - \pi_V)/P_0]$$

When correlating the normalized volume of carbon dioxide evolved per unit mass of aliquot samples with given calibrated mass percentage of sulfuric acid, the linear correlation exhibit a slope b and an intercept a corresponding a mass percentage of usually 0.1-0.5 wt.% H_2SO_4. This can be explained by the fact that a weak aqueous solution of sulfuric acid below that threshold does not possesses a pH high enough to decompose the carbonate thus the proper correlation is as follows:

$$A(\text{wt.\% } H_2SO_4) = a + b\, V_{CO2}(NTP)$$

The empirical slope and intercept coefficients must be determined experimentally for each equipment, and each prepared Kolbe's solutions, as they are impacted by the gas burette and the exact Bunsen's coefficient for the solution composition.

17 Conclusions

For 200 years, a plethora of sulfation techniques have been used worldwide across the mineral, metallurgical, and chemical industries. Some of these processes are implemented at a very large scale for the production of essential commodities such as superphosphate fertilizers, and titanium dioxide pigment while other sulfation processes are focussing on the production of small tonnages of high priced chemicals such as for instance lithium, beryllium, and rare earths.

Interestingly, in the last three decades, a certain number of novel and disruptive sulfation processes were devised and tested at the pilot scale by various private and public companies proposing novel routes to produce inorganic chemicals that are currently processed industrially using totally different routes (e.g., alkaline, hydrochloric).

Among them several of these new approaches are intended to address the recycling of industrial by-products, mineral processing wastes, and effluent that are currently not monetized, and thus simply disposed-off and landfilled due to the lack of suitable technologies.

As a general rule, all these processes, share three common denominators: (i) the low energy consumption due to the strong exothermic character of most sulfation reactions; (ii) the fast reaction kinetics as the operating temperature ranges of 180-325°C is much higher than the temperature range of 100-105°C encountered with diluted aqueous solutions of inorganic acids and bases; and (iii) the low operating cost due to the affordability of sulfuric acid especially at locations benefiting from a nearby smelter or an existing source of sulfur.

Nowadays sulfation processes occupies a key position but they also rely on the production, supply, and availability of sulfuric acid which originates itself from smelters roasting copper, nickel or zinc sulfide minerals, or from the burning of native sulfur or from sulfur by-produced from desulfurization methods.

Finally, any future attempts to replace pyrometallurgical roasting processes by novel hydrometallurgical processes will have to be assessed carefully as such changes will impact not only the commercial availability of sulfuric acid but also the industrial production of associated rare and minor metals and metalloids, (In, Ge, Sb, Se, Te, Re) and precious metals (Au, Ag).

18 Appendices

18.1 Acronyms

Table 88 – List of Acronyms

Acronyms	Description
AN	Acid number
APU	Acid retardation
ATH	Aluminum trihydrate
BPE	Boiling point elevation
BPR	Boiling point rise
BRGM	*Bureau de Recherches Géologiques et Minières*
CANMET	*Canada Center for Mineral and Energy Technology*
CAPEX	Capital expenditures
CAS No.	Chemical Abstract Service Registry Number
CBMM	*Companhia Brasileira de Metalurgia e Mineração*
CPVC	Chlorinated-polyvinylchloride
DECHEMA	*Deutsche Gesellschaft für chemisches Apparatewesen*
EH&S	Environmental, Health & Safety
EPDM	Ethylene propylene diene monomer
FSA	Fluorosilicic (Fluosilicic) acid
GHG	Greenhouse gases
HDPE	High density polyethylene
HSAL	Hot sulfuric acid leaching
HX	Heat exchanger
IR	Infrared electromagnetic radiation
mpy	mils per year
MVR	Mechanical vapor recompression
NAMB	National Materials Advisory Board
NMR	Nuclear magnetic resonance
O.V.	Oil of Vitriol

Acronyms	Description
OPEX	Operating cost
PE	Polyethylene
PFA	Perfluoroalkoxy
PHAR	Pickliq hydrochloric acid regeneration
PLS	Pregnant leach solution
PP	Polypropylene
PPI	Personal protective equipment
PTFE	Polytetrafluoroethylene
PVC	Polyvinylchloride
PVDF	Polyvinylidene fluoride
S.G.	Specific gravity
SI	*Système International d'Unités*
SOFREM	*Société Française d'Électrométallurgie*
USBM	*U.S. Bureau of Mines*
USCS	*U.S. Customary System*
US DOE	*U.S. Department of Energy*
US EPA	*U.S. Environmental Protection Agency*
VEPT	VanadiumCorp-Electrochem Process Technology

18.2 Latin and Greek Symbols

Table 89 – Latin symbols

Symbol	Physical Quantity	SI unit	USCS unit
$[R_D]$	Molar refraction	m³/mol	ft³/lb-mol
°Bé	Baume degree	none	none
c_P	Specific heat capacity (isobar)	J/kg/K	Btu/lb/°R
E	Nernst standard electrode potential	V/SHE	V/SHE
g	Specific Gibbs enthalpy	J/kg	Btu/lb
G	Molar Gibbs enthalpy	J/mol	Btu/mol
h	Specific enthalpy	J/kg	Btu/lb
H	Molar enthalpy	J/mol	Btu/mol
H_o	Hammett acidity function	none	none
k	Thermal conductivity	W/m/K	(Btu/h)/ft/°R

Symbol	Physical Quantity	SI unit	USCS unit
K_a	Acidity constant	none	none
k_D	Specific refractivity	m^3/kg	ft^3/lb
K_D	Specific refraction	m^3/kg	ft^3/lb
M	Molar mass	kg/kmol	lb/lb-mol
n_D	Refractive index	none	none
pH	Potential hydrogen	none	none
r_D	Molar refractivity	m^3/mol	ft^3/lb-mol
s	Specific entropy	J/kg/K	Btu/lb/°R
S	Molar entropy	J/mol/K	Btu/lb-mol/°R
T_b, *b.p.*	Boiling (vaporizing) point	K	°R
T_m, *m.p.*	Melting (freezing) point	K	°R
u	Sound velocity	m/s	ft/s
w_i	Mass fraction of species i	none	none

Table 90 – Greek symbols

Greek symbol	Physical Quantity	SI unit	USCS unit
ε	Dielectric permittivity	none	none
η	Dynamic (absolute) viscosity	Pa.s	lbf-s/ft^2
κ	Electrical conductivity	S/m	S/ft
π	Vapor pressure	Pa	lbf/ft^2
ρ	Mass density	kg/m^3	lb/ft^3
σ	Surface tension	N/m	lbf/ft

18.3 Mixing Rules

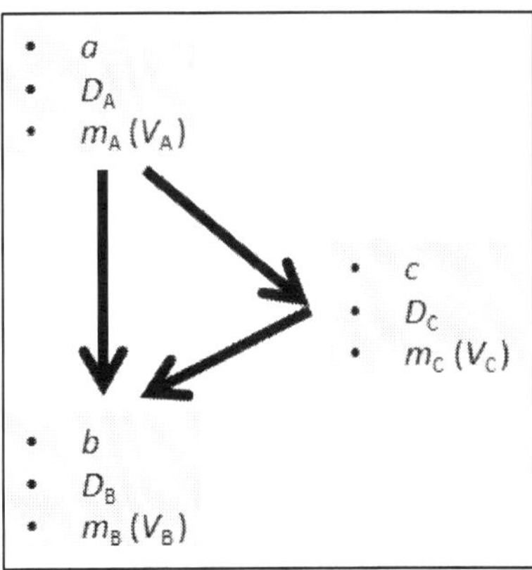

Figure 51 – Mixing rules

Table 91 – Physical quantities used in mixing rules

Physical quantity	Description
D_A	mass density of heavier solution (i.e., more concentrated)
D_B	mass density of lighter solution (i.e., less concentrated)
D_C	mass density of final prepared solution
a	mass fraction of H_2SO_4 in heavier solution (i.e., more concentrated)
b	mass fraction of H_2SO_4 in lighter solution (i.e., less concentrated)
c	mass fraction of H_2SO_4 in final prepared solution
$m_A (V_A)$	mass (resp. volume) of heavier solution (more concentrated)
$m_B (V_B)$	mass (resp. volume) of lighter solution (i.e., less concentrated)
$m_C (V_C)$	mass (resp. volume) of final prepared solution

Table 92 – Mixing rules

Masses	$m_C = m_A + m_B$ $c \cdot m_C = a \cdot m_A + b \cdot m_B$	$m_A = m_C \dfrac{(c - b)}{(a - b)}$	$m_B = m_C \dfrac{(a - c)}{(a - b)}$	$m_A = m_B \dfrac{(c - b)}{(a - c)}$
Volumes (*)	$V_C = V_A + V_B \pm \partial V_M$	$V_A = V_C \dfrac{(D_C - D_B)}{(D_A - D_B)}$	$V_B = V_C \dfrac{(D_A - D_C)}{(D_A - D_B)}$	$V_A = V_B \dfrac{(D_C - D_B)}{(D_A - D_C)}$
Densities	$D_C = \dfrac{1}{\left[\frac{a}{D_A} + \frac{b}{D_B} \pm \frac{dV_B}{m_C}\right]}$	$m_A = m_C \dfrac{D_A}{D_C} \dfrac{(D_C - D_B)}{(D_A - D_B)}$	$m_B = m_C \dfrac{D_B}{D_C} \dfrac{(D_A - D_C)}{(D_A - D_B)}$	$m_A = m_B \dfrac{D_A}{D_B} \dfrac{(D_C - D_B)}{(D_A - D_C)}$

Notes: (*) these equations are exact when the volume change upon mixing (∂V_M) is negligible with no contraction or expansion.

18.4 Glossary and Definitions

Sulfation. The sulfation denotes at large all the techniques in the chemical and metallurgical industries using sulfuric acid as main reactant with the intent to produce metal sulfates irrespective of the concentration of the sulfuric acid, the operating temperature, and the mode of heating.

Sulfuric acid cracking. The sulfuric acid cracking is a relatively vague term as sometimes it is used for describing the sulfuric acid digestion performed in the 180 220°C temperature range or on the opposite, the sulfation roasting conducted at high temperature in the vicinity of 350-400°C. Therefore, we tried in this monograph to avoid this terminology unless in specific cases where the term is well adopted such as in the case of the sulfation of the monazite.

Sulfuric acid digestion. The sulfuric acid digestion denotes a method of sulfation using strong concentrated sulfuric acid in order to decompose a solid to obtain either a sulfation cake, or a pasty solid or a solution. The mass ratio of sulfuric acid-to-solid must be sufficient to ensure that all the valuable metals within the solid yield soluble and insoluble metal sulfates while at the end of the digestion at least 80 percent of the initial mass of solids must be sulfated. Usually the temperature at which the sulfuric acid digestion is performed range from 80°C up to the boiling point of the sulfuric acid that of course varies with the initial concentration of sulfuric acid an can reach up to 340°C theoretically when using the azeotrope composition with 96 wt.% H_2SO_4 is reached. However for practical reasons (i.e., reaction kinetics, corrosion of materials, economics, and environment) the sulfuric acid digestion is conducted industrially at temperatures often below 250°C.

Sulfuric acid leaching. The term sulfuric acid leaching is a process where the solids are treated with dilute sulfuric acid below 40 wt.% H_2SO_4 to convert the valuable metals within the leached solid into soluble sulfate salts while the impurities and inert solids remains insoluble. Usually, because the dilution of the acid, the sulfuric leaching is performed under milder conditions than those used during the digestion usually between 80°C and 120°C. However, to address the temperature limitation and improve the dissolution reaction kinetics, some processes perform the sulfuric acid leaching under pressure inside acid resistant autoclaves and other types of pressure vessels.

Sulfation baking. The sulfation baking denotes a sulfation technique consisting of sulfating a mixture of concentrated sulfuric acid or oleum impregnating a divided solid at a temperature ranging from 150°C to 340°C.

Sulfation roasting. The sulfation roasting denotes a high temperature sulfation technique consisting of roasting a mixture of concentrated sulfuric acid or oleum impregnating a divided solid in air at a temperature ranging from 340°C and 700°C. Conversely, it also includes the sulfation of a solid with alkali-metal sulfates, bisulfates, and pyrosulfates in the solid or molten state.

18.5 Prices of Chemicals

The average prices during the period 2018-2022 for the chemicals and commodities often used by the sulfation process, produced or by-produced as a result of sulfation processes are reported in Table 93.

Table 93 – Prices of chemicals (2018-2022)

Chemical name, grade, purity	Chemical formula (IUPAC)	Price ($/tonne)	Price ($/lb.)
Ammonia (anhydrous), tanks	NH_3	$520.00	0.2359
Ammonium sulfate, bulk, f.o.b.	$(NH_4)_2SO_4$	$266.00	0.1207
Ferric sulfate, bags	$Fe_2(SO_4)_3$	$300.00	0.1361
Ferrous sulfate heptahydrate, drums	$FeSO_4.7H_2O$	$150.00	0.0680
Gypsum (crude, 95 wt.%)	$CaSO_4.2H_2O$	$9.00	0.0041
Hydrochloric acid (35-37 wt.%), tanks	HCl	$165.00	0.0748
Hydrofluoric acid (60 wt.%), tanks, f.o.b.	HF	$800.00	0.3629
Nitric acid (69 wt.%)	HNO_3	$490.00	0.2223
Oleum (65 wt.%), tanks	$H_2S_2O_7 + SO_3$	$165.00	0.0748
Phosphoric acid (85 wt.%), tanks	H_3PO_4	$713.00	0.3234
Potassium sulfate (Fertilizer) (90-95 wt.%)	K_2SO_4	$700.00	0.3175
Sodium sulfate (anhydrous)	Na_2SO_4	$200.00	0.0907
Sulfur (flour), bulk, f.o.b.	S_8	$90.00	0.0408
Sulfur dioxide (liquid)	SO_2	$1,450.00	0.6577
Sulfuric acid (93 wt.%), tanks	H_2SO_4	$55.00	0.0249
Sulfuric acid (96 wt.%), tanks	H_2SO_4	$150.00	0.0680
Titanium dioxide (98 wt.%), bagged, f.o.b.	TiO_2	$3,569.00	1.619

19 Index

About the Author

Dr. François Cardarelli, President and Owner of the Canadian company *Electrochem Technologies & Materials Inc.*, is an industrial chemist with a strong physical-chemistry background and a doctorate in chemical engineering from the University Paul Sabatier (UPS) Toulouse III, and he is the inventor and co-inventor of 14 patents, and the sole author of three reference handbooks published worldwide by Springer since 1996.

He has over 32 years of industrial experience in North America and Europe in developing electrochemical, chemical, and metallurgical processes for winning, refining or producing a variety of metals, alloys, and inorganic chemicals either from aqueous solutions or molten salts media.

A particular area of his professional expertise is the chemical, and electrochemical processing of mining residues, metallurgical wastes, and industrial effluents, the manufacture of industrial electrodes, the electrochemical production of vanadium electrolyte, the pyro- and hydrometallurgical production of vanadium, niobium, tantalum, and tungsten compounds and chemicals, and finally the manufacture of novel industrial materials. All these processes are covered by patents enforced in many jurisdictions.

Dr. François Cardarelli is a member in good standing of the following professional organizations and societies: *American Institute of Chemical Engineers* (AIChE)[Lifetime member], *American Chemical Society* (ACS), *Chemical Institute of Canada* (CIC), *Canadian Society for Chemical Engineering* (CSChE), *The Electrochemical Society* (ECS), *Mineralogical Society of America* (MSA), *Ordre des Chimistes du Québec* (OCQ), The *Oughtred Society* (OS), and *The Minerals, Metals and Materials Society* (TMS).

Made in the USA
Columbia, SC
06 January 2023

74467558R00165